How to "Ace" Statistics 101: Textbook

Print Edition

Anthony Rodriguez
Zetetic Library
Guttenberg, New Jersey

Copyright / Disclaimer

Author / Publisher

Author

The author of this book is Dr. Anthony Rodriguez. After fleeing Cuba as a teenager in the early 60s, Dr. Rodriguez studied in the United States, worked in Corporate America, and became a successful entrepreneur. Most recently, he taught undergraduate courses at the University of Phoenix in Jersey City and at LIM College in Manhattan. Nowadays, he is engaged in research and writing about myriad topics. As a business person, he traveled all over the world; as a private pilot, he circled the Big Apple; and, as a scuba diver, he eye-balled sharks in the Florida Keys.

Publisher

The publisher of this book is the Zetetic Library —www.ZeteticLibrary.com— a virtual bookstore that offers bargain-priced, user-friendly books about myriad topics. The Zetetic Library's contact information is:

Dr. Anthony Rodriguez
7002 Boulevard East, Suite 39-F
Guttenberg, New Jersey 07093
(800) 290-8850 (Toll free) or (201) 854-8800
DrAnthonyRodriguez@ZeteticLibrary.com

Table of Contents

FRONT MATTER

TEXT

DESCRIPTIVE STATISTICS

PROBABILITY

NON-PARAMETRIC TESTS

BACK MATTER

Preface

Do you want to get an A+ in Statistics 101? This book will show you how! The 221-page, softcover book is a bargain-priced, user-friendly textbook about descriptive statistics, probability, and inferential statistics (including non-parametric testing).

The book is divided into four parts:

- Descriptive Statistics
- Probability
- Inferential Statistics
- Non-Parametric Tests

Unlike today's textbooks, the book displays no superfluous glitz, just useful information. And, for the examples, the book includes calculations by hand.

Although I tried my best, please email any errors or omissions to:

DrAnthonyRodriguez@ZeteticLibrary.com

Introduction to Statistics

Statistics is all about data and information. Statistics are ways of turning raw data into meaningful information.

Data are characteristics of people, animals, or things. For example:

- In the United States, the average height of adult males is about 5 feet, 10 inches.
- Worldwide, the color of about 25% of new automobiles is white.

Information is organized data. For example, Table 1.1 and Figure 1.2 on Page 2 are examples of organized data; that is, examples of information.

Branches of Statistics

The study of statistics is usually divided into three parts: descriptive statistics, probability, and inferential statistics. Probability is the foundation of Inferential statistics.

Descriptive Statistics

Descriptive statistics are ways of collecting, summarizing, analyzing, and presenting data.

Data may be described with numbers. For example, Table 1.1 shows the results of the millions of votes cast in the 2012 federal elections.

Table 1.1				
Affiliation	**Senate**		**House**	
	#	**%**	**#**	**%**
Democrat	53	53.0	201	46.2
Republican	45	45.0	234	53.8
Independent	2	2.0	0	0.0
Total	100	100.0	435	100.0

Data may also be described with graphs. For example, Figure 1.1 on Page 2 shows the colors of 20 automobiles.

Figure 1.1

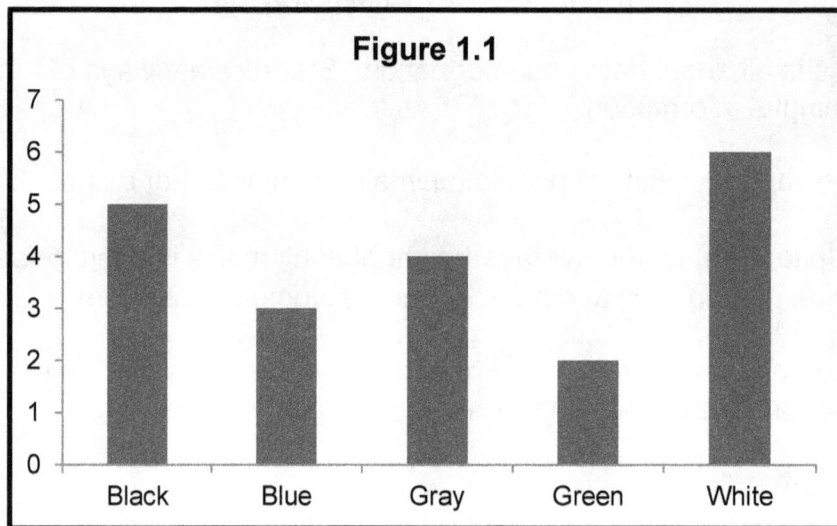

Probability

Probability is the chance of something happening —or not happening— in the long run. For example, the probability of getting a head in a coin flip is 1 out of 2, or 1/2, or 0.50, or 50%.

Populations and Samples

A population is all the members of a group. For example, all the students of a college is a population. Populations may be subsets of other populations. For example, a population made up of all the female students of a college is a subset of the population made up of all the students of the college. Both are well-defined populations. That is, a population is whatever one defines the population to be.

A sample is a portion of the population. For example, 1/4 of the female students of the college is a sample of the population made up of all the female students of the college.

A simple random sample (SRS) is a sample drawn from a population in which every member of the population has an equal chance of being drawn. For example, one may draw names from a bowl.

Other types of samples, for example, sampling only one's friends —a type of sample called convenience samples— are not truly representative of populations.

Inferential Statistics

Inferential statistics are ways of drawing conclusions about the characteristics of populations. In inferential statistics, oftentimes, one uses the characteristics of random samples drawn from the populations rather than using the characteristics of whole populations. Inferential statistics is divided into two parts: estimation and hypothesis testing.

Estimation

Statistics are ways of estimating the characteristics of populations based on the characteristics of random samples drawn from the populations. For example, one may estimate the average weight of the U.S. population by weighing a portion of the population.

Hypothesis Testing

Statistics are ways of comparing the characteristics of populations based on the characteristics of random samples drawn from the populations. For example, one may compare the average weight of the U.S. male population to the average weight of the U.S. female population by weighing a portion of each population.

A hypothesis is a statement about a characteristic of a population. For example, one may say that the average weight of the U.S. male population is equal to the average weight of the U.S. female population. Hypothesis testing are ways of failing to reject —that is, accepting— or rejecting such statements.

Another Meaning of Statistics

Statistics are also values that describe the characteristics of samples whereas parameters are values that describe the characteristics of populations. For example, the average height of 1/3 of the male students of a college is a statistic of the sample whereas the average height of all the male students of the college is a parameter of the population.

Oftentimes, the parameters of populations are unknown and the statistics of samples are used to estimate the parameters of populations. For example, the average weight of a portion of the male students of a college may be used to estimate the average weight of all the male students of the college.

Blank Page

Collecting Data

Statistics begins with data collection. Recall that, oftentimes, one collects data from random samples drawn from populations rather than from whole populations. Data is collected by observation or by experimentation. Oneself may collect the data or one may use data collected by others.

Variables

In statistics, the characteristics of people, animals, or things are called variables. For example, people's height is a variable. Variables can take different values. For example, a person's height may be 5 feet, 5-1/2 feet, 6 feet, or whatever.

Observational Studies

In observational studies, one simply records the values of the variables being studied. For example, one may record the gender, height, and weight of the students at a college.

Experimental Studies

In experimental studies, one manipulates the values of one variable —called the independent variable— and, then, measures the change on the values of another variable —called the dependent variable. For example, as shown in Figure 2.1, one may subject high school seniors to varying levels of math tutoring —the independent variable— and, then, measure the students' math ability —the dependent variable.

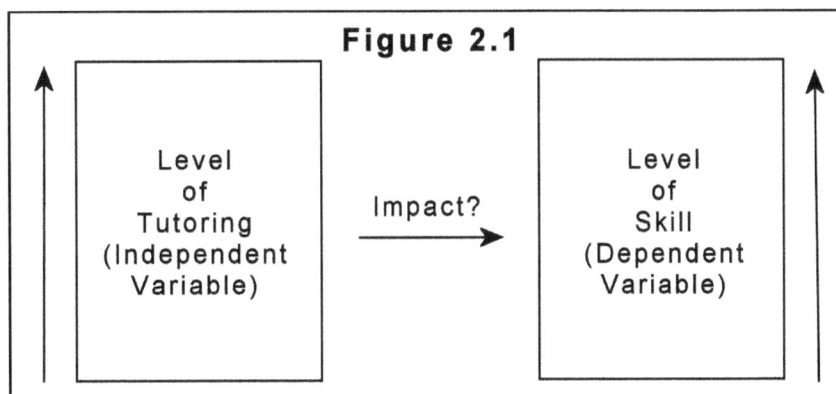

Figure 2.1

Level of Tutoring (Independent Variable) → Impact? → Level of Skill (Dependent Variable)

Primary Data versus Secondary Data

Oneself may collect data. Such data is called primary data. Or, one may use data collected by others. Such data is called secondary data. For example, one may use data collected by the U.S. Census Bureau.

Data, Variables, and Constants

Recall that statistics is all about data. Also recall that data are characteristics of people, animals, or things. Finally recall that, in statistics, the characteristics of something are called variables. There are different types of data and different types of variables.

Univariate Data

Data that is made up of one variable is called univariate. For example, data that shows people's height is univariate.

Bivariate Data

Data that is made up of two variables is called bivariate. For example, data that shows people's weight and height is bivariate.

Multivariate Data

Data that is made up of more than two variables is called multivariate. For example, data that shows people's gender, weight, and height is multivariate.

Random Variables

Variables can take different values. That is, the values of variables are random. For example, the days of the week may be Sunday, Monday, Tuesday, or whatever.

Qualitative Variables

Qualitative variables, also called categorical variables or nominal variables, are characteristics that can be described with words. There is no hierarchy between the values of the variables. That is, there is no relationship such as "higher than" between the values. For example, the color of an automobile is a qualitative variable. The color of an automobile may be white, black, gray, or whatever.

Dichotomous Variables

Dichotomous variables are qualitative variables that can take one of only two values. For example, a person's gender is a dichotomous variable. A person's gender can only be male or female.

Ordinal Variables

Ordinal variables are qualitative variables that can be ordered. There is a hierarchy between the values of the variables. That is, there is a relationship such as "higher than" between the values. For example, the rank of a soldier is an ordinal variable. The rank of a soldier may be corporal, sergeant, lieutenant, or whatever.

Quantitative Variables

A quantitative variable is a characteristic that can be described with numbers. For example, a person's height is a quantitative variable. A person's height may be 5 feet, 5-1/2 feet, 6 feet, or whatever.

Discrete Variables

Discrete variables are quantitative variables that can take only a finite number of values. For example, the size of men's shoes is a discrete variable. The size of men's shoes may be 6, 6-1/2, 7, or whatever.

Continuous Variables

Continuous variables are quantitative variables that can take an infinite number of values. For example, a person's weight is a continuous variable. A person's weight may be 150 pounds, 150.5 pounds, 150.55 pounds, or whatever.

Independent Variables

In experimental studies, independent variables are variables that are manipulated by the researcher. For example, the researcher may subject high school seniors to varying levels of math tutoring.

Dependent Variables

In experimental studies, dependent variables are variables that are influenced by changes in the independent variables. For example, the researcher may measure the math ability of the high school seniors who were subjected to varying levels of math tutoring.

Constants

Constants are values that do not change. For example, the value of pi, 3.14159..., is a constant.

Scales of Measurement

Recall that variables are characteristics of something that can take different values. Measuring are ways of assigning values to variables based on scales. That is, measuring are ways of sorting, ordering, or counting the values of variables. According to Stevens, there are four types of scales of measurement: nominal, ordinal, interval, and ratio. [1]

Nominal Scales

Nominal scales use words to sort nominal variables. There is no hierarchy between the values of the variables. That is, there is no relationship such as "higher than" between the values. For example, a scale that shows the color of automobiles is a nominal scale. The color of automobiles may be white, black, gray, or whatever.

Ordinal Scales

Ordinal scales also use words to order ordinal variables. There is a hierarchy between the values of the variables. That is, there is a relationship such as "higher than" between the values. For example, a scale that shows the status of college students is an ordinal scale. The status of college students may be freshman, sophomore, junior, or senior.

Interval Scales

Interval scales use numbers to count quantitative variables. Interval scales are constructed to display equal intervals without a true zero. Zeroes do not mean the absence of the characteristic; zeroes are simply points of the scale. For example, the Celsius and the Fahrenheit temperature scales are interval scales (See Figure 4.1).

Figure 4.1						
Celcius	30		20		10	0
Difference		10		10		10
Fahrenheit	86		68		50	32
Difference		18		18		18

Looking at both scales simultaneously (See Figure 4.1), you can say that 30 degrees Celsius is 20 degrees higher than 10 degrees Celsius and 86 degrees Fahrenheit is 36 degrees higher than 50 degrees Fahrenheit. But, you cannot say that 30 degrees Celsius is three times as hot as 10 degrees Celsius because 86 degrees Fahrenheit is not three times as hot as 50 degrees Fahrenheit. That is, you can add and subtract the values but you can not multiply or divide the values.

Ratio Scales

Ratio scales also use numbers to count quantitative variables. Ratio scales are constructed to display equal intervals with a true zero. Zeroes do mean the absence of the characteristic. For example, ordinary rulers are ratio scales (See Figure 4.2).

Figure 4.2						
Inches	3		2		1	0
Difference		1		1		1
Centimeters	7.62		5.08		2.54	0.00
Difference		2.54		2.54		2.54

Looking at both scales simultaneously (See Figure 4.2), you can say that 3 inches is 2 inches longer than 1 inch and 7.62 centimeters is 5.08 centimeters longer than 2.54 centimeters. And, you can also say that 3 inches is three times as long as 1 inch because 7.62 centimeters is three times as long as 2.54 centimeters. That is, you can add, subtract, multiply, and divide the values.

Properties of Scales

Table 4.1 summarizes the properties of the scales of measurement.

Table 4.1				
	Nominal	**Ordinal**	**Interval**	**Ratio**
Difference	X	X	X	X
Hierarchy		X	X	X
Magnitude			X	X
Zero				X
+/-			X	X
x/÷				X

Endnote

[1] Stevens, S.S. (1946). On the theory of scales of measurement. *Science, 103*(2684), 677-680.

Distributions

Recall that a variable is a characteristic of something that can take different values. A distribution is a set —that is, a list— of the values of a variable. For example. Table 5.1 shows the distribution of the colors of 20 automobiles. And, Table 5.2 shows the distribution of the weights of 25 male college students.

Table 5.1				
Color				
White	Blue	Black	Gray	White
Gray	White	Green	Black	Gray
Blue	Black	White	Green	Black
Black	Gray	White	White	Blue

Table 5.2				
Weight (Pounds)				
125	118	140	147	175
116	119	137	156	146
102	133	169	107	145
134	113	176	192	193
190	200	171	157	171

Discrete Distributions

Recall that discrete variables are quantitative variables that can take only a finite number of values. Discrete distributions show the distributions of discrete variables.

Continuous Distributions

Recall that continuous variables are quantitative variables that can take an infinite number of values. Continuous distributions show the distributions of continuous variables.

Frequency Distributions

Frequency means how many or how often. A frequency distribution, also called a frequency table. is a summary of the values of a variable that shows the absolute frequencies and/or the relative frequencies —that is, the percentages— of the values. For example, Table 5.3 shows the absolute frequencies and the relative frequencies of the colors of the 20 automobiles.

Table 5.3		
Color	**Absolute Frequency**	**Relative Frequency (%)**
Black	5	25.0
Blue	3	15.0
Gray	4	20.0
Green	2	10.0
White	6	30.0
Total	20	100.0

Oftentimes, the values of the variable are grouped. For example, Table 5.4 shows the grouped absolute frequencies, the grouped relative frequencies, and the grouped cumulative frequencies of the weights of the 25 male college students.

Table 5.4			
Weight (Pounds)	**Absolute Frequency**	**Relative Frequency (%)**	**Cumulative Frequency (%)**
100 to 124	6	24.0	24.0
125 to 149	8	32.0	56.0
150 to174	5	20.0	76.0
175 to 200	6	24.0	100.0
Total	25	100.0	

Bar Charts

A bar chart is a graph of the frequency distribution of a qualitative variable. For example, Figure 5.1 on Page 13 shows a bar chart of the absolute frequencies of the colors of the 20 automobiles.

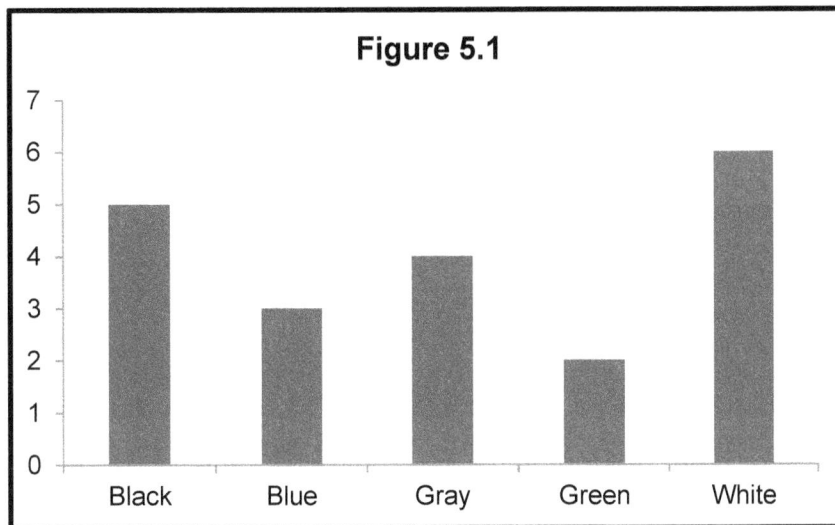

Figure 5.1

Pie Charts

A pie chart is another graph of the frequency distribution of a qualitative variable. For example, Figure 5.2 shows a pie chart of the absolute frequencies of the colors of the 20 automobiles.

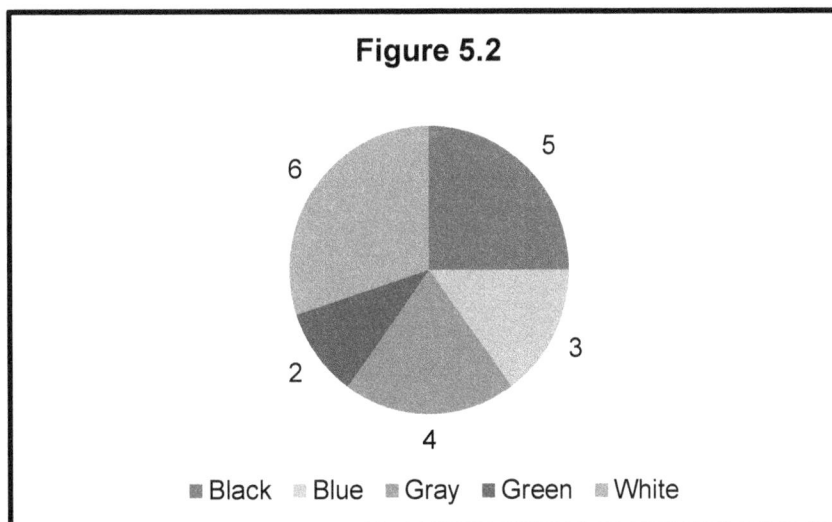

Figure 5.2

Frequency Histograms

A frequency histogram is a graph of the frequency distribution of a discrete variable. The area of the histogram equals the frequency of the values. For example, Figure 5.3 on Page 14 shows a frequency histogram of the grouped absolute frequencies of the weights of the 25 male college students.

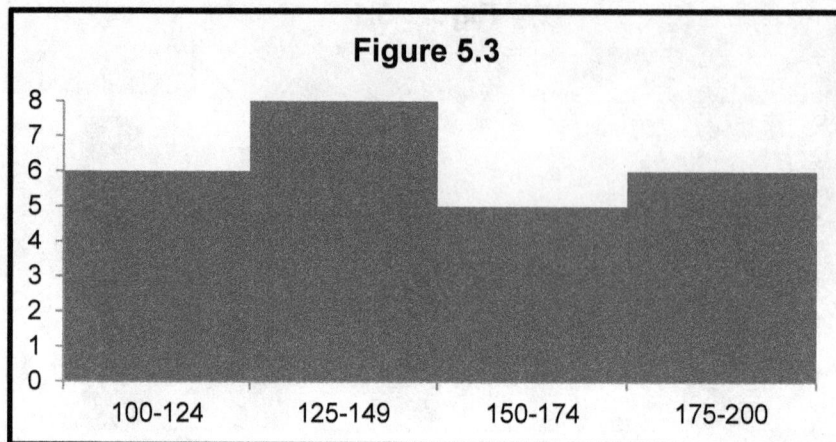

Figure 5.3

Frequency Polygons

A frequency polygon is another graph of the frequency distribution of a quantitative variable. For example, Figure 5.4 shows a frequency polygon of the weights of the 25 male college students.

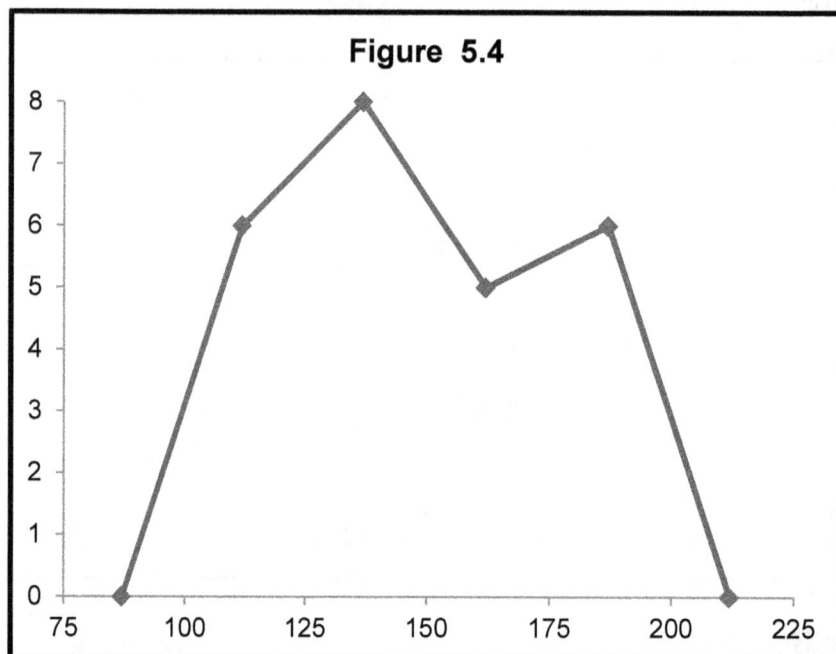

Figure 5.4

Density Curves

A density curve is a "smoothed" histogram (See Figure 5.5 on Page 15). A density curve is a graph of the frequency distribution of a continuous variable. The area under the curve equals the frequency of the values.

Figure 5.5

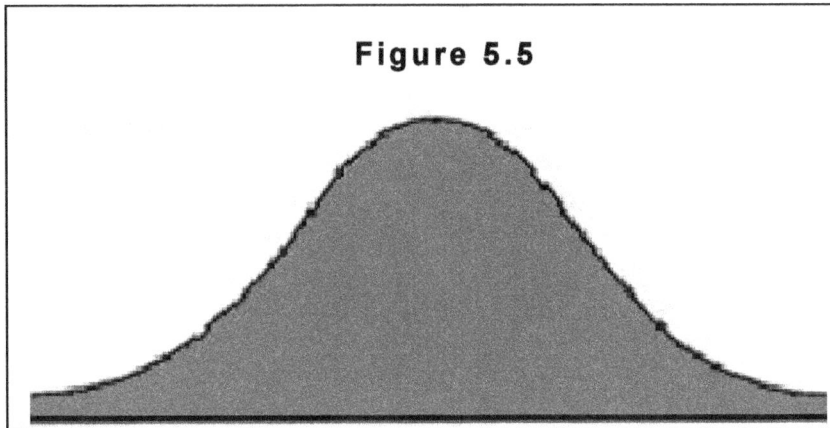

Distributions versus Lists

Recall that, a distribution is a set —that is, a list— of the values of a variable. Statisticians use the word "distribution" instead of the word "list" because "distribution" implies some sort of order. For example, Figure 5.6 shows a scatterplot of the distribution of the weights of the 25 male college students.

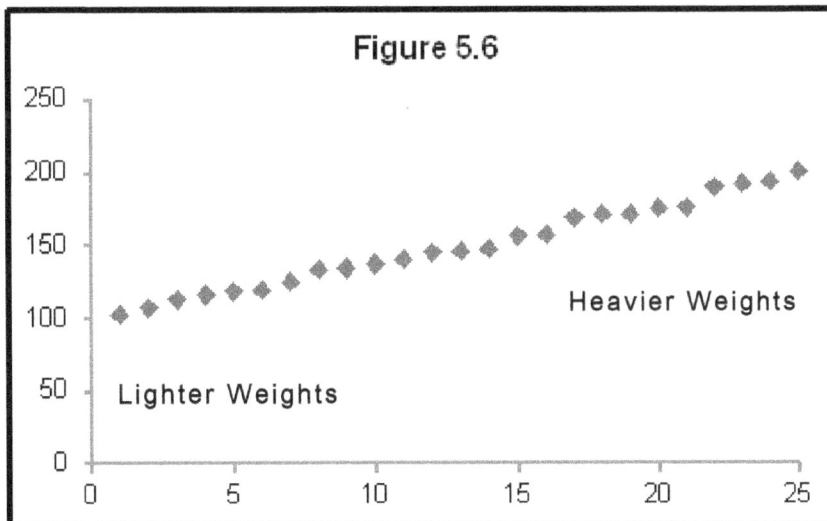

Figure 5.6

Characteristics of Distributions

Figure 5.7 on Page 16 shows some of the characteristics of distributions: the center, the spread, and the shape, for example, "bell-curved". These characteristics are described in later chapters.

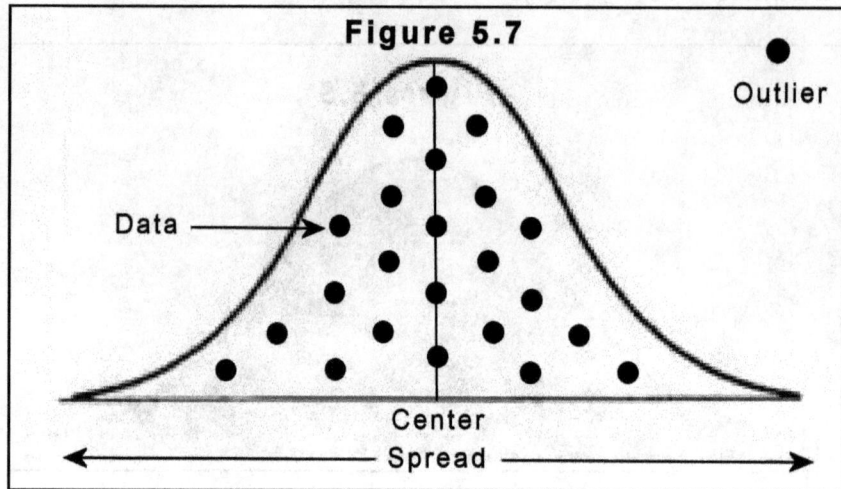

Figure 5.7

Measures of Central Tendency

Recall that a distribution is a set —that is, a list— of the values of a variable. Measures of central tendencies are single values —that is, statistics of samples or parameters of populations— that describe the middle value of a distribution. The most common measure of central tendency of qualitative variables is the mode. And, the most common measures of central tendency of quantitative variables are the mode, the mean, and the median.

Sample Data

Table 6.1 shows the absolute frequencies of the colors of 20 automobiles. And, Table 6.2 shows the distribution of the weights of 25 male college students, which, for the purposes of this chapter, may be a population or a sample.

Table 6.1	
Color	**Absolute Frequency**
Black	5
Blue	3
Gray	4
Green	2
White	6
Total	20

Table 6.2				
Weight (Pounds)				
125	118	140	147	175
116	119	137	156	146
102	133	169	107	145
134	113	176	192	193
190	200	171	157	171

Table 6.3 on Page 18 shows the 2014 estimated U.S. population of U.S.-born and of foreign-born. [1] And, Table 6.4 on Page 18 shows the number of U.S.-born and of foreign-born in a random sample of 1,000 Americans.

Table 6.3 (Millions)	
U.S.-born	276.4
Foreign-born	42.3
Total	318.7

Table 6.4	
U.S.-born	870
Foreign-born	130
Total	1,000

Mode

The mode is the most frequent value of a distribution. For example, the mode of the absolute frequencies shown in Table 6.1 is white, which appears six times. And, the mode of the distribution shown in Table 6.2 is 171, which appears twice.

Mean

The mean is the arithmetic average of the values of a quantitative distribution. The formula for the population mean, μ, is ...

$$\mu = \frac{\sum X}{N}$$

... where μ is the population mean, X are the population values and N is the population size.

And, the formula for the sample mean, x-bar, is ...

$$\bar{x} = \frac{\sum x}{n}$$

... where x-bar is the sample mean, x are the sample values and n is the sample size.

For example, the mean of the distribution shown in Table 6.2 on Page 17 —which may be the distribution of a population or the distribution of a sample— is 149.28 and is calculated by adding all the values and, then, dividing the result by the number of values:

$$(125+116+...+171)/25=149.28$$

Mean of Proportions

A proportion is the ratio of a portion to the whole. The formula for the population proportion, p, is …

$$p=\frac{X}{N}$$

… where p is the population proportion, X is the number —that is, the frequency — of values that satisfy a condition, and N is the population size.

Based on the formula for the mean of a binomial distribution, which is described in a later chapter, the mean of the population proportion, μ_p, is equal to p:

$$\mu_p=p$$

For example, the mean of the population proportion of foreign-born shown in Table 6.3 on Page 18 is 13.3%:

$$42.3/318.7=13.3\%$$

The formula for the sample proportion, p-hat, is …

$$p\text{-hat}=\frac{x}{n}$$

… where p-hat is the sample proportion, x is the number —that is, the frequency — of values that satisfy a condition, and n is the sample size.

Like the mean of the population proportion, the mean of the sample proportion, $\mu_{p\text{-hat}}$, is equal to p-hat:

$$\mu_{p\text{-hat}}=p\text{-hat}$$

For example, the mean of the sample proportion of foreign-born shown in Table 6.4 on Page 18 is 13.0%:

$$130/1{,}000 = 13.0\%$$

Median

The median is the middle value of an ordered distribution. For example, the median of the distribution shown in Table 6.2 on Page 17 is 146 and is found by ordering the distribution from low to high and, then, finding the middle value:

$$102, 107, \ldots, 146, \ldots, 193, 200$$

For even-numbered distributions, the median is the arithmetic average of the two middle values.

Endnote

[1] U.S. Census Bureau. (2015). *Projections of the size and composition of the U.S. population: 2014 to 2017*. Washington, DC: Author.

Measures of Dispersion

Recall that a distribution is a set —that is, a list— of the values of a variable. Measures of dispersion are single values —that is, statistics of samples or parameters of populations— that describe the variability —that is, the spread— of the values of a distribution. The most common measures of dispersion of quantitative variables are the range, the variance, and the standard deviation.

Sample Data

Table 7.1 shows the distribution of the weights of 25 male college students, which, for the purposes of this chapter, may be a population or a sample.

Table 7.1				
Weight (Pounds)				
125	118	140	147	175
116	119	137	156	146
102	133	169	107	145
134	113	176	192	193
190	200	171	157	171

Table 7.2 shows the 2014 estimated U.S. population of U.S.-born and of foreign-born. [1] And, Table 7.3 shows the number of U.S.-born and of foreign-born in a random sample of 1,000 Americans.

Table 7.2 (Millions)	
U.S.-born	276.4
Foreign-born	42.3
Total	318.7

Table 7.3	
U.S.-born	870
Foreign-born	130
Total	1,000

Range

The range shows the minimum value and the maximum value of a distribution. For example, the range of the distribution shown in Table 7.1 on Page 21 is 102-200.

Variance

A deviate is the difference between a value and the mean. The population variance is the average of the square of the deviations of the population values from the population mean. The formula for the population variance, σ^2, is ...

$$\sigma^2 = \frac{\sum (X-\mu)^2}{N}$$

... where σ^2 is the population variance, X are the population values, μ is the population mean, and N is the population size.

For example, the population variance of the distribution shown in Table 7.1 on Page 21 —if the distribution is a population— is 819.2416 and is calculated by adding the squares of the deviations of the population values from the population mean, which, as shown in Chapter 6, is 149.28, and, then, dividing the result by the population size:

$$((125-149.28)^2+(116-149.28)^2+...+(171-149.28)^2)/25=819.2416$$

And, the sample variance is the average of the square of the deviations of the sample values from the sample mean. The formula for the sample variance, s^2, is ...

$$s^2 = \frac{\sum (x-\bar{x})^2}{n-1}$$

... where s^2 is the sample variance, x are the sample values, x-bar is the sample mean, and n is the sample size.

For example, the sample variance of the distribution shown in Table 7.1 on Page 21 —if the distribution is a sample— is 853.3767 and is calculated by adding the squares of the deviations of the sample values from the sample mean, which, as shown in Chapter 6, is 149.28, and, then, dividing the result by the sample size minus one:

$$((125-149.28)^2+(116-149.28)^2+...+(171-149.28)^2)/24=853.3767$$

Variance of Proportions

Recall that a proportion is the ratio of a portion to the whole. The formula for the variance of the population proportion, σ^2_p, is …

$$\sigma^2_p = p * q$$

… where σ^2_p is the variance of the population proportion, p is the population proportion, and q is 1-p.

For example, the mean of the population proportion of foreign-born shown in Table 7.2 on Page 21 is 13.3% (See Chapter 6). Therefore, the variance of the population proportion is 0.1153%:

$$0.133*(1-0.133)=0.1153$$

And, the formula for the variance of the sample proportion, $\sigma^2_{p\text{-hat}}$, is …

$$\sigma^2_{p\text{-hat}} = p\text{-hat} * q\text{-hat}$$

… where $\sigma^2_{p\text{-hat}}$ is the variance of the sample proportion, p-hat is the sample proportion, and q-hat is 1 - p-hat.

For example, the mean of the sample proportion of foreign-born shown in Table 7.3 on Page 21 is 13.0% (See Chapter 6). Therefore, the variance of the sample proportion is 0.1131:

$$0.13*(1-0.13)=0.1131$$

Note that, for proportions, the population size and sample size are ignored.

Standard Deviation

The standard deviation is simply the square root of the variance. For example, the population standard deviation, σ, of the distribution shown in Table 7.1 on Page 21 —if the distribution is a population— is 28.6224:

$$SQRT(819.2416)=28.6224$$

And, the sample standard deviation, s, of the distribution shown in Table 7.1 on Page 21 —if the distribution is a sample— is 29.2126:

$$SQRT(853.3767)=29.2126$$

Standard Deviation of Proportions

The standard deviation is simply the square root of the variance. For example, the standard deviation of the population proportion, σ_p, of the above-referenced foreign-born is 0.3396:

$$SQRT(0.1153)=0.3396$$

And, the standard deviation of the sample proportion, $\sigma_{p\text{-hat}}$, of the above-referenced foreign-born is 0.3363:

$$SQRT(0.1131)=0.3363$$

Other Measures of Dispersion

The variance and the standard deviation measure the variability —that is, the spread— of the values of distributions. The standard error measures the variability — that is, the spread— of statistics of samples. Standard errors are described in later chapters.

Endnote

[1] U.S. Census Bureau. (2015). *Projections of the size and composition of the U.S. population: 2014 to 2017*. Washington, DC: Author.

Measures of Position

Recall that a distribution is a set —that is, a list— of the values of a variable. Measures of position measure the position of the values of an ordered distribution relative to the position of other values. Besides the median, the most common measures of position of quantitative variables are the percentiles, the deciles, and the quartiles.

Percentiles

Percentiles divide the values of an ordered distribution into 100 equal parts. According to some, but not to all, a percentile is the value which includes a percentage of all of the values. And, the percentile index is the number of values which includes a percentage of all of the values. The formula for the percentile index is ...

$$\text{Percentile Index} = k * \frac{p}{100}$$

...where k is the number of values and p is the percentile.

For example, Figure 8.1 shows a percentile index of 5. 5 is 4.8 (which is 6*80/100) rounded to 5. 5 is the number of values which includes 80% of all of the values and 11, which is the fifth value, is the 80th percentile.

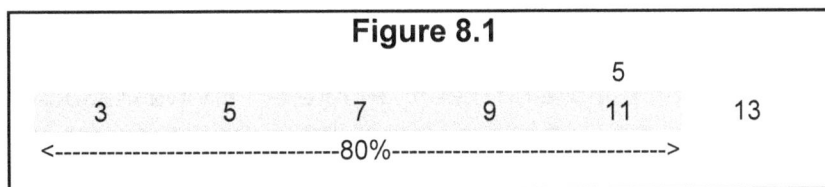

Figure 8.1					
				5	
3	5	7	9	11	13
<------------------------------80%------------------------------>					

Caveats! First, percentiles are values, not percentages, and, second, there are other ways of calculating percentiles.

Deciles

Deciles divide the values of an ordered distribution into ten equal parts. That is, deciles divide the values into ten groups of 10% each. Deciles are closely related to percentiles. For example, the 8th decile is equal to the 80th percentile.

Quartiles

Quartiles divide the values of an ordered distribution into four equal parts. That is, quartiles divide the values into four groups of 25% each. The second quartile equals the median.

Box Plots

Box plots are graphs that show the range and the quartiles (See Figure 8.2). Note that the second quartile equals the median.

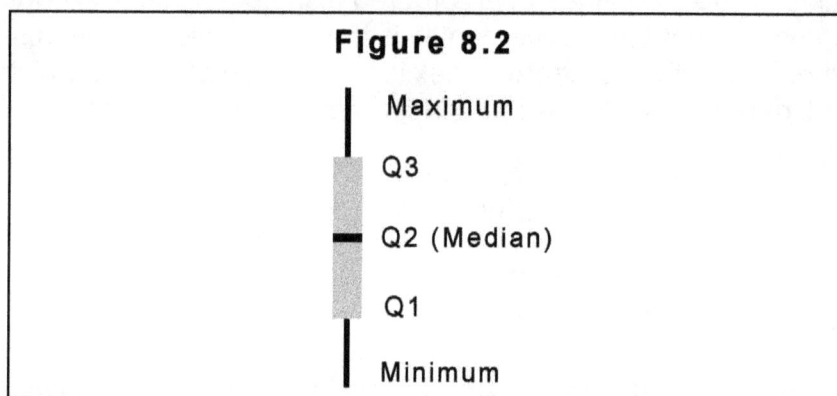

Figure 8.2

Maximum

Q3

Q2 (Median)

Q1

Minimum

Other Measures of Location

Other measures may be viewed as measures of location. For example, z scores show how many standard deviations a population value of a normally distributed population is away from the population mean. Figure 8.3 shows the location of the sample mean, x-bar, being way out to the right of the population mean, μ.

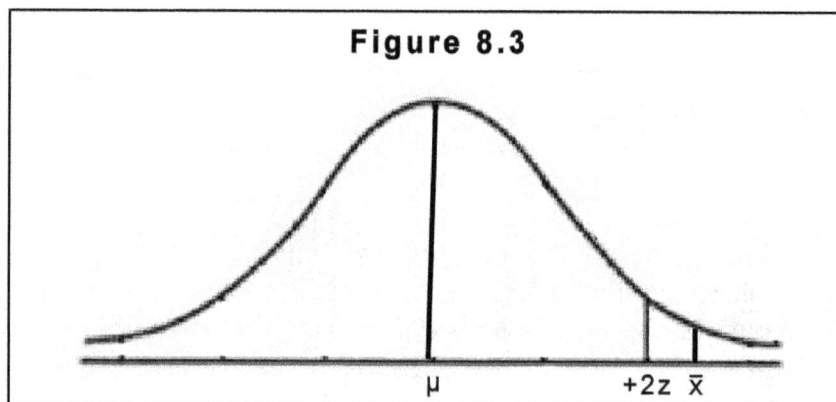

Figure 8.3

μ   +2z $\bar{x}$

Other Measures

Other measures include rates, ratios, percentages, and proportions.

Rates

Rates describe values in terms of other values. For example, an automobile may be traveling at 60 miles per hour, which is the rate of speed.

Ratios

Ratios describe the relationship between the frequencies of two values. For example, if there are 10 females and 20 males, the ratio of females to males is 1 to 2, 1:2, or 1/2.

Ratios also describe the relationship between the frequencies of values and the whole. For example, if there are 10 females and 20 males, the ratio of females is 1 to 3, 1:3, or 1/3. Recall that, in statistics, these ratios are also called proportions.

Percentages

Percentages also describe the relationship between the frequencies of values and the whole where the whole are hundreds. That is, percentages are frequencies per hundred. For example, if there are 10 females and 20 males, the percentage of females is 33-1/3. That is, 10/30*100=0.33...=33-1/3%.

Proportions

In mathematics, proportions state that two ratios are equal. That is, $f_1/f_2=f_3/f_4$. And, $f_1*f_4=f_2*f_3$. For example, one may say that, if a student reads two books in three days, then the student may read six books in nine days. That is, 2/3=6/9. And, 2*9=3*6.

Blank Page

Shapes of Distributions

Recall that frequency histograms and density curves are graphs of the frequency distributions of quantitative variables and that the areas of frequency histograms and density curves equal the frequency of the values. For the purposes of this chapter, histograms are used as examples.

Histograms show not only the absolute frequencies or the relative frequencies of the values of a variable but also the central tendency of the values and the variability — that is, the spread— of the values. That is, histograms show the "shapes" of distributions.

Skewness

Some distributions are symmetrical —that is, not skewed. That is, the histogram's left side mirrors the histogram's right side (See Figure 10.1). In symmetrical distributions, both the mean and the median are centered.

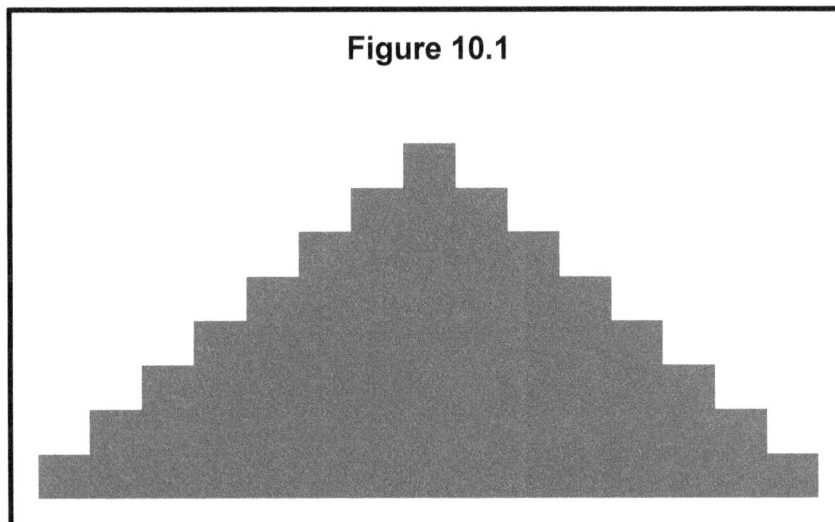

Figure 10.1

Some distributions are right-skewed —that is, positively skewed. That is, the histogram shows a tail on the right (See Figure 10.2 on Page 30). In positively-skewed distributions, the median is to the right of the mean.

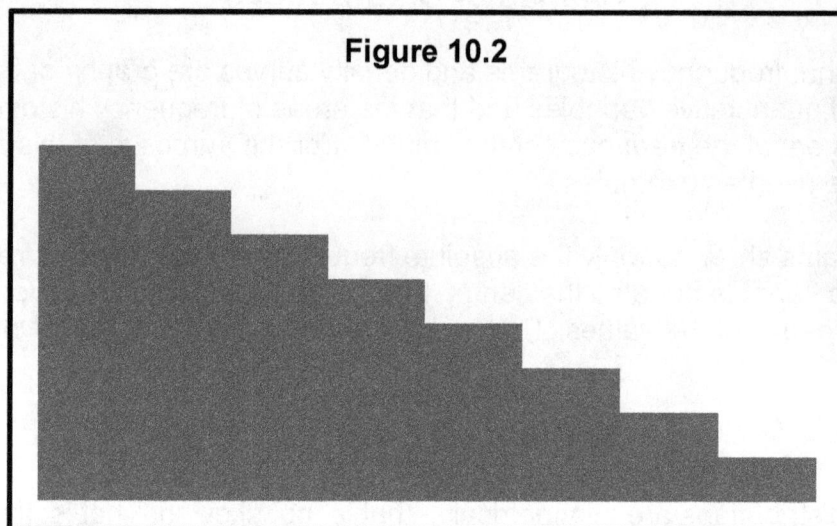

Figure 10.2

And, some distributions are left-skewed —that is, negatively skewed. That is, the histogram shows a tail on the left (See Figure 10.3). In negatively-skewed distributions, the median is to the left of the mean.

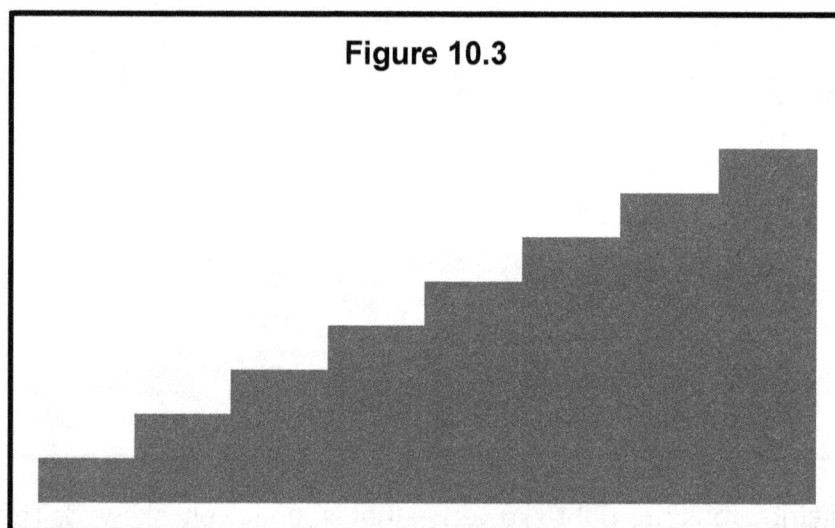

Figure 10.3

Kurtosis

The histograms of leptokurtic distributions show that the values are close together (See Figure 10.4 on Page 31).

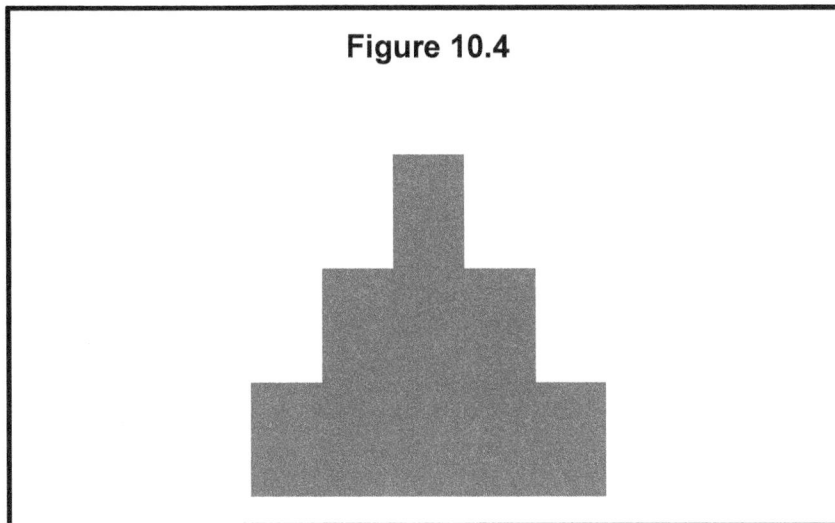

Figure 10.4

The histograms of platykurtic distributions show that the values are far apart (See Figure 10.5).

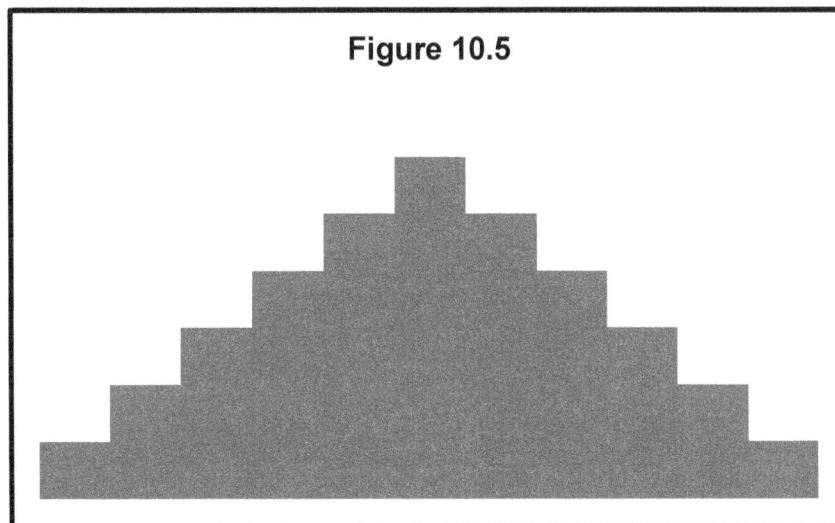

Figure 10.5

And, the histograms of mesokurtic distributions show that the values are "bell-shaped" (See Figure 10.6 on Page 32).

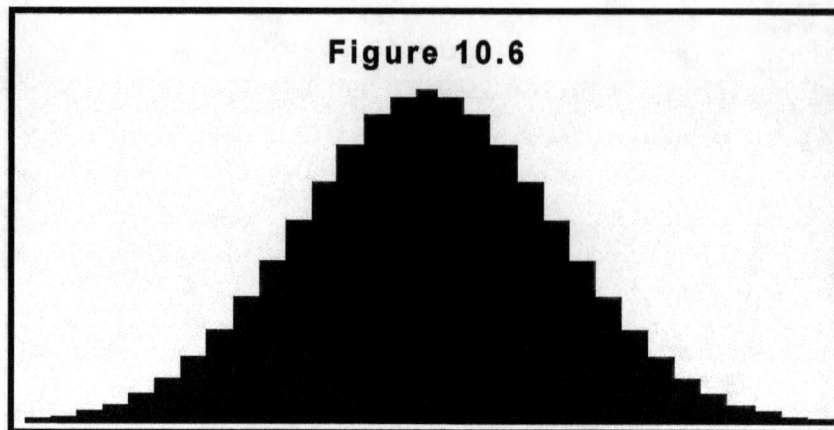

Figure 10.6

Modality

Some distributions are unimodal. That is, there is one peak (See Figure 10.6). Some distributions and bimodal. That is, there are two peaks (See Figure 10.7). And, some distributions are multimodal. That is, there are three or more peaks.

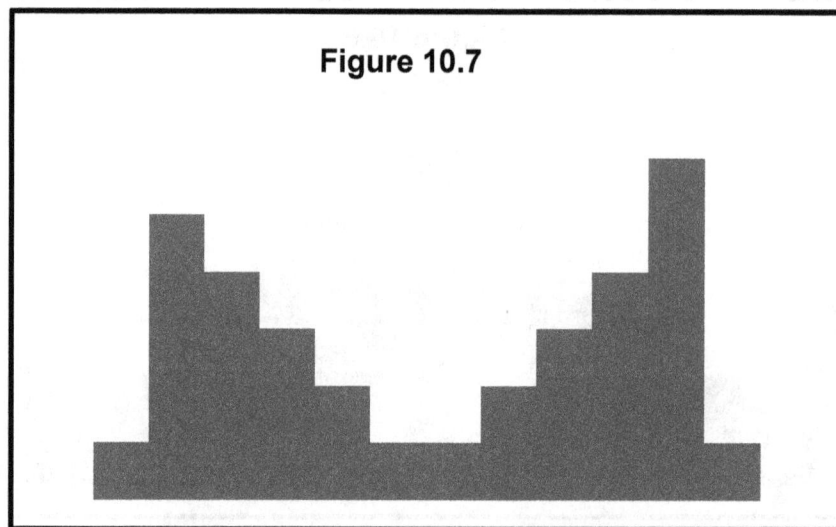

Figure 10.7

Correlation

The mean, the variance, and the standard deviation are measures —that is, parameters of populations or statistics of samples— of a single variable. Correlation describes the relationship between two variables. For example, in the real world, people's height and weight are related to each other. That is, generally speaking, taller people weigh more. Correlation measures the relationship between the variables; correlation does not measure the changes of the variables.

Types

Two variables are positively correlated when, as the value of one variable increases, the value of the other variable also increases. For example, people's height and weight are positively correlated. Two variables are negatively correlated when, as the value of one variable increases, the value of the other variable decreases. For example, drinking alcohol and driving ability are negatively correlated. And, two variables are not correlated when, as the value of one variable increases, the value of the other variable either increases or decreases. For example, people's height and math ability are not correlated.

Sample Data

Table 11.1 shows the distribution of the height and weight of 25 male college students.

Table 11.1				
Height (Inches) and Weight (Pounds)				
62.4, 102	65.7, 119	67.8, 140	68.3, 157	72.3, 176
62.6, 107	66.1, 125	67.9, 145	68.4, 169	72.6, 190
63.5, 113	66.7, 133	67.9, 146	68.7, 171	72.8, 192
63.9, 116	66.9, 134	68.0, 147	69.4, 171	73.0, 193
65.2, 118	67.3, 137	68.1, 156	69.7, 175	74.2, 200

Scatterplots

Scatterplots are graphs that show the relationship between two variables. Figure 11.1 on Page 34 shows the scatterplot for the sample data shown in Table 11.1. The upward trend shows that there is a positive correlation between height and weight.

Figure 11.1

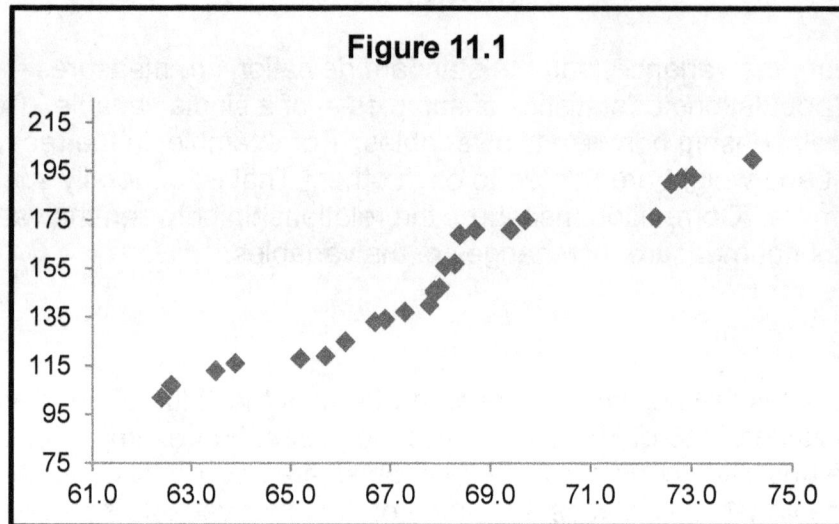

Pearson's Correlation Coefficient

The correlation coefficient measures the relationship between two variables. The formula for Pearson's correlation coefficient, r, is ...

$$r=\frac{n\sum xy-\sum x\sum y}{\sqrt{[n\sum x^2-(\sum x)^2][n\sum y^2-(\sum y)^2]}}$$

... where r is the correlation coefficient, n is the number of pairs, and x and y are the values.

Calculation

This section shows the steps to be followed for calculating the correlation coefficient of the sample data shown in Table 11.1 on Page 33. The correlation coefficient is 0.9670:

First. Calculate the the sum of the products of x and y:

(62.4*102)+(62.6*107)+...+(74.2*200)=255,872.3

Second. Calculate the product of the sum of x and the sum of y:

(62.4+62.6+...+74.2)*(102+107+...+200)=6,342,160.8

Third. Calculate the sum of the squares of x:

$$(62.4)^2+(62.6)^2+...+(74.2)^2=115,767.90$$

... and the sum of the squares of y:

$$(102)^2+(107)^2+...+(200)^2=577,594$$

Fourth. Calculate the square of the sum of x:

$$(62.4+62.6+...+74.2)^2=2,887,960.36$$

... and the square of the sum of y:

$$(102+107+...+200)^2=13,927,824$$

Fifth. Calculate Pearson's correlation coefficient:

$$(25*255,872.3-6,342,160.8)/SQRT((25*115,767.90-2,887,960.36)(25*577,594-13,927,824))=0.9670$$

Coefficient of Determination

The coefficient of determination measures the percentage of the variability of the variables that is due to the relationship between the variables. The coefficient of determination, r^2, is simply the square of the correlation coefficient, r. For example, the coefficient of determination for the sample data shown in Table 11.1 on Page 33 is 0.9351:

$$(0.9670)^2=0.9351$$

That is, 94% of the variability of the variables is due to the relationship between the variables.

Caveat!

Correlation does not imply causation! For example, in large U.S. cities, there are both a large number of churches and a high number of crimes. Yet, one cannot say that churches cause crimes, or vice versa.

Blank Page

Simple Regression

Recall that correlation measures the relationship between two variables. Simple regression also describes the relationship between two variables. But, regression measures the change in the independent variable and the corresponding average change in the dependent variable.

For example, recall that, in the real world, people's height and weight are related to each other. That is, generally speaking, taller people weigh more. Regression measures the change in height —which is the independent variable— and the corresponding average change in weight —which is the dependent variable.

Sample Data

Table 12.1 shows the distribution of the height and weight of 25 male college students. And, Figure 12.1 shows the corresponding scatterplot.

Table 12.1				
Height (Inches) and Weights (Pounds)				
62.4, 102	65.7, 119	67.8, 140	68.3, 157	72.3, 176
62.6, 107	66.1, 125	67.9, 145	68.4, 169	72.6, 190
63.5, 113	66.7, 133	67.9, 146	68.7, 171	72.8, 192
63.9, 116	66.9, 134	68.0, 147	69.4, 171	73.0, 193
65.2, 118	67.3, 137	68.1, 156	69.7, 175	74.2, 200

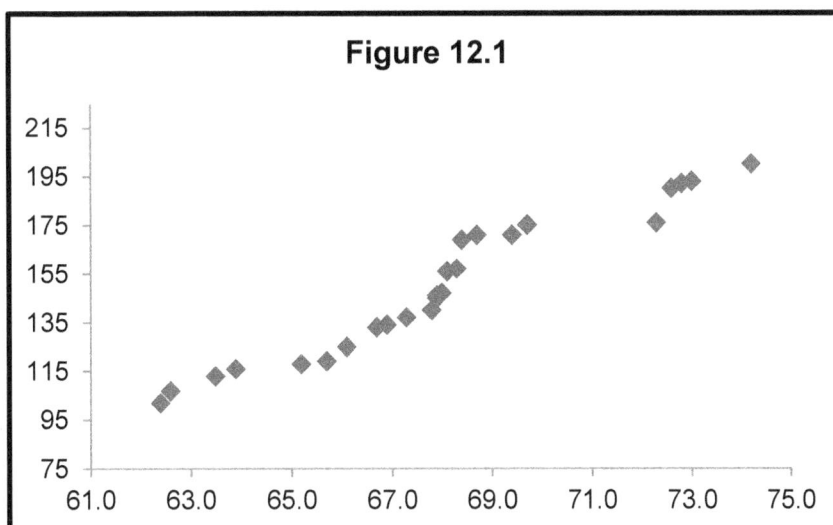

Figure 12.1

Regression Line

The regression line shows the relationship between the variables (See Figure 12.2).

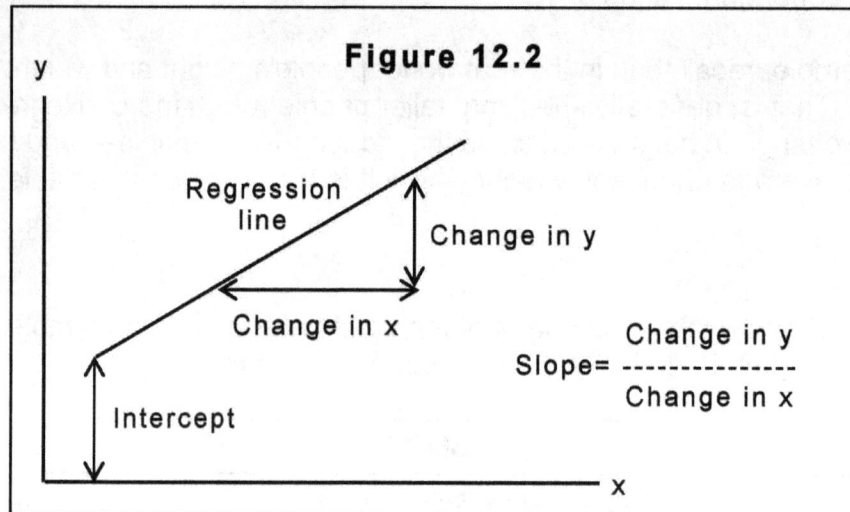

Figure 12.2

Formulas

The formula for the regression line is ...

$$y=a+bx$$

... where y is the dependent variable, a is the intercept, b is the slope of the line, also called the regression coefficient, and x is the independent variable. The slope of the line is the change in y —that is, the rise— divided by the change in x —that is, the run.

The formula for the slope, b, is ...

$$b=\frac{n\sum xy-(\sum x)(\sum y)}{n\sum x^2-(\sum x)^2}$$

... where b is the slope, n is the number of pairs, x is the independent variable, and y is the dependent variable.

And, the formula for the intercept, a, is ...

$$a = \frac{\sum y - b\left(\sum x\right)}{n}$$

… where a is the intercept, y is the dependent variable, b is the slope, x is the dependent variable, and n is the number of pairs.

Calculation

This section shows the steps to be followed for calculating the slope, b, of the sample data shown in Table 12.1 on Page 37. The slope, b, is 8.7615:

First. Calculate the sum of the products of x —the independent variable— and y —the dependent variable:

$$(62.4*102) + (62.6*107) + \ldots + (74.2*200) = 255{,}872.3$$

Second. Calculate the sum of x:

$$62.4 + 62.6 + \ldots + 74.2 = 1{,}699.4$$

… and the sum of y:

$$102 + 107 + \ldots + 200 = 3{,}732$$

Third. Calculate the sum of the squares of x:

$$(62.4)^2 + (62.6)^2 + \ldots + (74.2)^2 = 115{,}767.90$$

Fourth. Calculate the slope:

$$(25*255{,}872.3 - (1{,}699.4*3{,}732)) / ((25*115{,}767.90) - (1{,}699.4)^2) = 8.7615$$

This section shows the steps to be followed for calculating the intercept, a, of the sample data shown in Table 12.1 on Page 37. The intercept, a, is -446.29:

First. Calculate the sum of y:

$$102 + 107 + \ldots + 200 = 3{,}732$$

... and the sum of x:

$$62.4+62.6+...+74.2=1,699.4$$

Second. Calculate the intercept:

$$(3,732-(8.7615*1,699.4))/25=-446.29$$

Introduction to Probability

Recall that probability is the chance of something happening —or not happening — in the long run.   For example, the probability of getting a head in a coin flip is 1 out of 2, or 1/2, or 0.50, or 50%. And, the probability of not getting a head is also 1 out of 2, or 1/2, or 0.50, or 50%. The probability of getting a one in a die roll is 1 out of 6, or 1/6, or 16.666..., or 16-2/3%. And, the probability of not getting a one in a die roll is 5 out of 6, or 5/6, or 83.333..., or 83-1/3%.

Trial

Flipping a coin is a trial. And, rolling a die is also a trial.

Experiment

Repeated trials are called an experiment. For example, flipping a coin five times is an experiment. And, rolling a die ten times is also an experiment.

Outcome

An outcome is the result of a trial. For example, getting a head is an outcome of flipping a coin. And, getting a one is an outcome of rolling a die.

Sample Space

A sample space is a set of all possible outcomes. For example, the sample space of flipping a coin is a head and a tail. The sample space of rolling a die is a one, a two, a three, a four, a five, and a six. Table 13.1 shows the sample space of flipping two coins.

Table 13.1	
First Coin	**Second Coin**
Head	Head
Head	Tail
Tail	Head
Tail	Tail

Event

An event is a set of outcomes that satisfy a condition. That is, an event is a subset of the sample space. For example, getting exactly two heads after flipping two coins is an event (See Table 13.2).

Table 13.2		
First Coin	**Second Coin**	**Probability**
Head	Head	1/4
Head	Tail	1/4
Tail	Head	1/4
Tail	Tail	1/4
Total		4/4

Independent Events

Two events are independent if the occurrence of the events are not related. For example, the outcomes, heads or tails, of flipping two coins are independent of each other.

Mutually Exclusive Events

Two events are mutually exclusive if the occurrence of the events cannot happen together. For example, the outcomes, a head or a tail, of flipping one coin are mutually exclusive.

Venn Diagrams

Oftentimes, Venn diagrams are used to graphically display probabilities. For example, Figure 13.1 on Page 43 shows a sample space S, an event A, an event B, and the intersection —that is, the overlap— of events A and B.

Figure 13.1

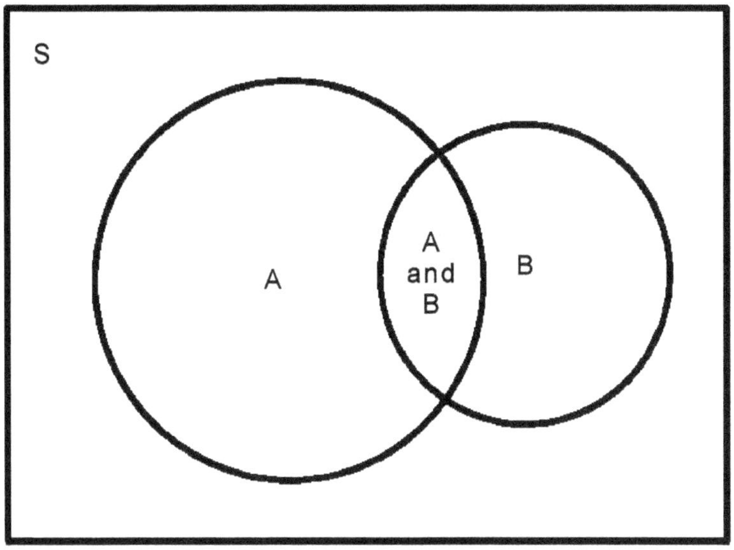

Probability of an Event

The probability of an event, P(X), is …

$$P(X) = \frac{r}{n}$$

… where P(X) is the probability of the event, r is the number of outcomes that satisfy the condition and n is the total number of outcomes –that is, the sample space.

For example, as Table 13.2 on Page 42 shows, the probability of getting two heads after flipping two coins is 1 out of 4, or 1/4, or 0.25, or 25%. Also note that the sum of all probabilities is 4 out of 4, or 4/4, or 1.0, or 100%.

Notation

In probability, "∩"" means "and", "U" means "or", "|" means "given that", and "‾" means "not".

Multiplication Rule

For two independent events, the probability of the two events occurring together is equal to the product of their probabilities. That is...

$$P(A \cap B) = P(A) * P(B|A)$$

For example, as Table 13.2 on Page 42 shows, the probability of getting two heads after flipping two coins is 1/4:

$$1/2 * 1/2 = 1/4$$

Addition Rule

For two mutually exclusive events, the probability of one of the two events occurring is equal to the sum of their probabilities. That is...

$$P(A \cup B) = P(A) + P(B)$$

For example, the probability of getting a one or a six in a die roll is 1/3:

$$1/6 + 1/6 = 1/3$$

And, for any two events, the probability of one of the two events occurring is equal to the sum of their probabilities less the probability of the two events occurring together. That is ...

$$P(A \cup B) = P(A) + P(B) - P(A \cap B)$$

For example, the probability of drawing a club or an ace from a deck of cards is 4/13 (See Figure 13.2 on Page 45):

$$13/52 + 4/52 - (13/52 * 4/52) =$$
$$17/52 - 1/52 = 4/13$$

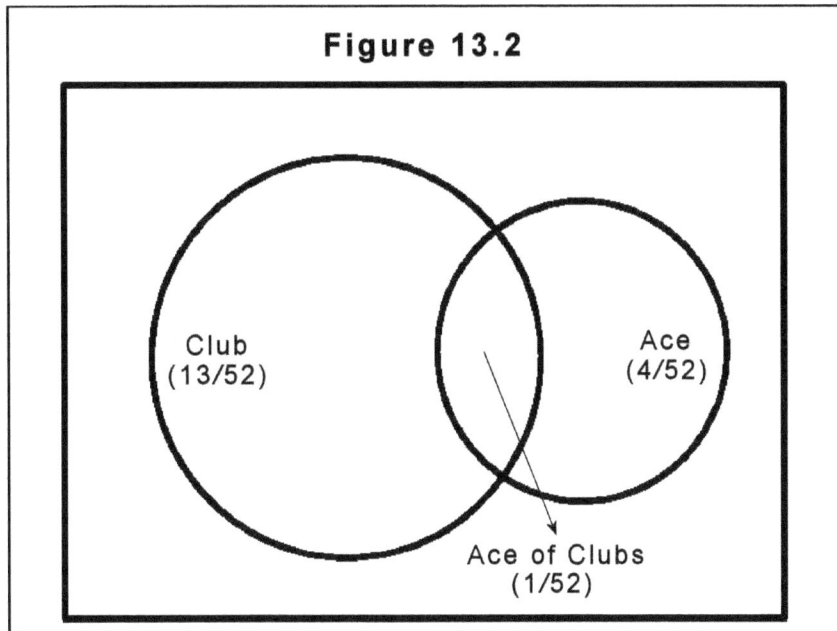

Figure 13.2

Club
(13/52)

Ace
(4/52)

Ace of Clubs
(1/52)

Conditional Probability

Conditional probability is the probability that an event (Event B) will occur, given that another event (Event A) has already occurred. That is ...

$$P(B|A) = \frac{P(B \cap A)}{P(A)} = \frac{P(B) * P(A)}{P(A)}$$

For example, say one draws a card from a deck of cards. What is the probability that the card drawn is an ace (Event B) if the card drawn is a heart (Event A)? The probability is 1/13:

$$(4/52 * 13/52)/(13/52) = 1/13$$

As another example, say one draws two cards from a deck of cards without replacement. What is the probability that the second card drawn is a heart (Event B) if the first card drawn is a heart (Event A)? The probability is 12/51:

$$(12/51 * 13/52)/(13/52) = 12/51$$

Bayes' Theorem

Bayes' Theorem states that ...

$$P(B|A)=\frac{P(B)*P(A|B)}{(P(B)*P(A|B))+(P(\bar{B})*P(A|\bar{B}))}$$

For example, according to Froelic and Stephenson [1], if 45% of a random sample of students are males, 55% are females, 39% of males are blue-eyed, and 33% of females are blue-eyed, then, the probability that a blue-eyed student (Event A) is female (Event B) is 51%:

$$(0.55*0.33)/((0.55*0.33)+(0.45*0.39))=0.51$$

Endnote

[1] Froelic, A.G. & Stephenson, W.R. (2013). Does eye color depend on gender? It might depend on who or how your ask. *Journal of Statistics Education, 21*(2), 1-10.

Probability Distributions

Recall that probability is the chance of something happening —or not happening — in the long run. Also recall that a distribution is a set —that is, a list— of the values of a variable. A probability distribution is a distribution of the probabilities of the values of a variable.

Intuitively, the chances of getting a head or a tail in a coin flip are equal. That is, in the theoretical world, the probability of getting a head or a tail in a coin flip is 1 out of 2, or 1/2, or 0.50, or 50%. However, in the real world —where the values of variables are random— hundreds, or even thousands, of coin flips will most likely not yield exactly 50% heads and 50% tails.

A model is a representation of reality. For example, a photograph is a two-dimensional model. And, a statue is a three-dimensional model. Probability distributions reflect the theoretical world, not the real world. That is, probability distributions are mathematical models.

For example, Table 14.1 shows the sample space of a dice roll. And, Table 14.2 on Page 48 and Figure 14.1 on Page 48 show the theoretical probabilities of the sum of the outcomes of an infinite number of dice rolls.

In the real world —where the values of variables are random— hundreds, or even thousands, of dice rolls will most likely not yield exactly twelve every 36 times. However, in the theoretical world, that is, in the world of probability distributions, the probability of twelve is 1/36 (See Table 14.2 on Page 48 and Figure 14.1 on Page 48).

Table 14.1					
1,1	2,1	3,1	4,1	5,1	6,1
1,2	2,2	3,2	4,2	5,2	6,2
1,3	2,3	3,3	4,3	5,3	6,3
1,4	2,4	3,4	4,4	5,4	6,4
1,5	2,5	3,5	4,5	5,5	6,5
1,6	2,6	3,6	4,6	5,6	6,6

Table 14.2	
X **(Sum)**	**Probability**
2	1/36
3	2/36
4	3/36
5	4/36
6	5/36
7	6/37
8	5/36
9	4/36
10	3/36
11	2/36
12	1/36
Total	36/36

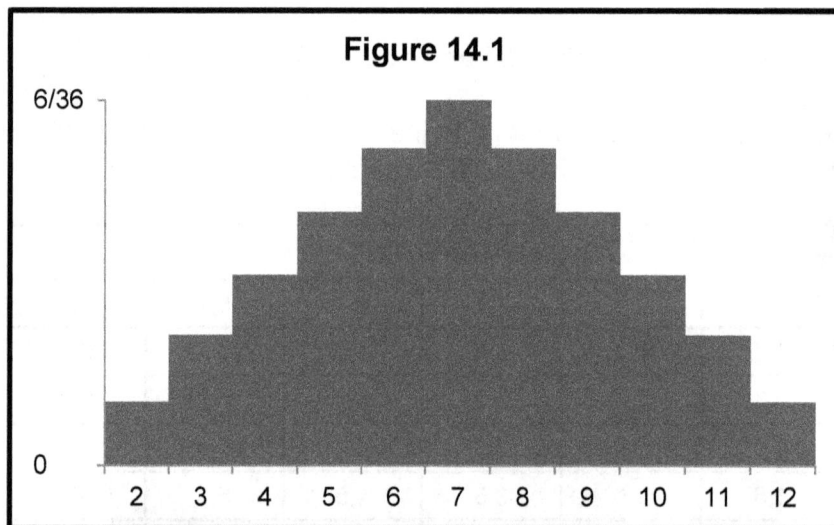

Figure 14.1

Table 14.3 on Page 49 and Figure 14.2 on Page 49 show the theoretical probabilities of an infinite, uniformly-distributed number of fractions between 0 and 5.

In the real world —where the values of variables are random— hundreds, or even thousands, of times of drawing a fraction between 0 and 5 will most likely not yield exactly a fraction between 4 and 5 every five times. However, in the theoretical world, that is, in the world of probability distributions, the probability of a fraction between 4 and 5 is 1/5 (See Table 14.3 and Figure 14.2).

Table 14.3	
X	**Probability**
0 to 1	1/5
1 to 2	1/5
2 to 3	1/5
3 to 4	1/5
4 to 5	1/5
Total	5/5

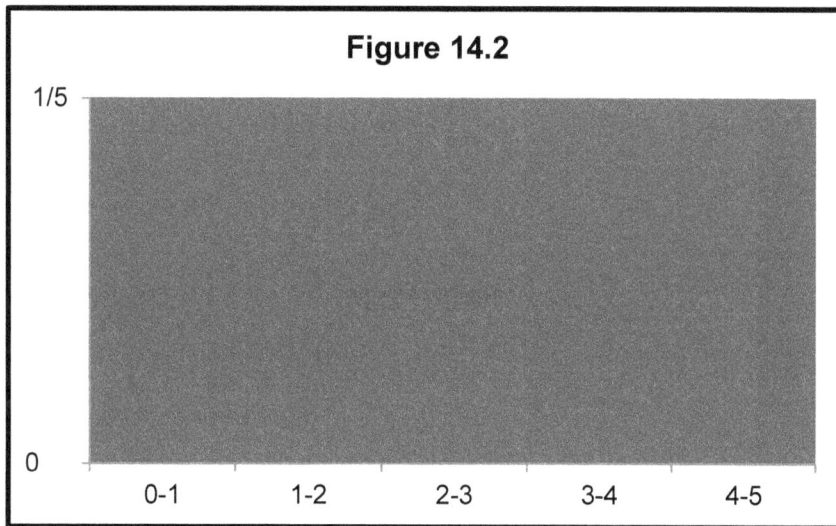

Figure 14.2

Blank Page

Discrete Probability Distributions

Recall that a discrete distribution shows the distribution of a discrete variable. Also recall that a probability distribution is a distribution of the theoretical probabilities of the values of a variable. A discrete probability distribution is a distribution of the theoretical probabilities of a discrete variable. For example, Table 15.1 shows the theoretical probabilities of the sum of the outcomes of an infinite number of dice rolls.

Table 15.1	
X **(Sum)**	**Probability**
2	1/36
3	2/36
4	3/36
5	4/36
6	5/36
7	6/37
8	5/36
9	4/36
10	3/36
11	2/36
12	1/36
Total	36/36

Graphs

Recall that a frequency histogram is a graph of the frequency distribution of a discrete variable and that the area of the histogram equals the frequency of the values. Discrete probability distributions are graphed as frequency histograms and the area equals one. For example, Figure 15.1 on Page 52 shows the frequency histogram of the theoretical probabilities of the sum of the outcomes of an infinite number of dice rolls.

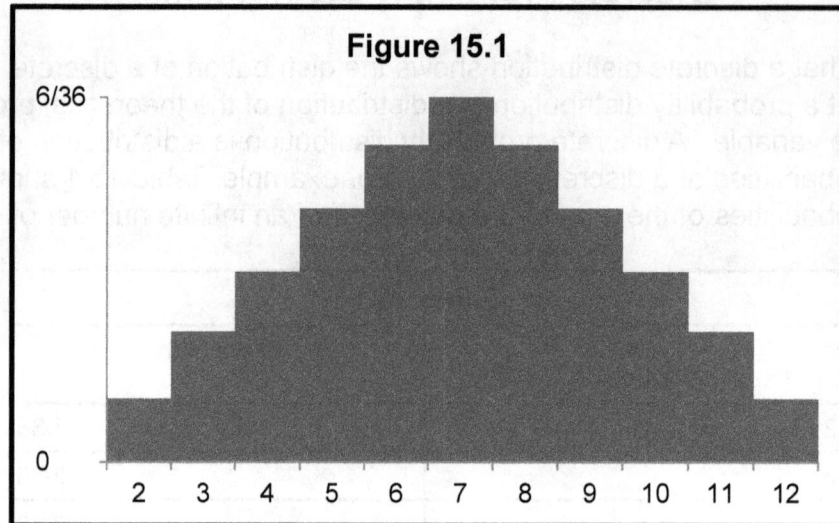

Figure 15.1

Mean

Recall that the mean is a measure of central tendency. The formula for the mean of a discrete probability distribution, E(X), is ...

$$E(X)=\sum X * P(X)$$

... where E(X) is the mean, X are the outcomes and P(X) are the probabilities of the outcomes.

For example, the mean of the distribution shown in Table 15.1 on Page 51 is 7:

$$(2*1/36)+(3*2/36)+...+(12*1/36)=7$$

Variance

Recall that the variance is a measure of dispersion. The formula for the variance of a discrete probability distribution, VAR(X), is ...

$$VAR(X)=\sum \left[X-E(X)\right]^{2} * P(X)$$

... where VAR(X) is the variance, X are the outcomes, E(X) is the mean of the distribution, and P(X) are the probabilities of the outcomes.

For example, the variance of the distribution shown in Table 15.1 on Page 51 is 5.8333:

$$((2-7)^2*(1/36))+((3-7)^2)*(2/36))+...+((12-7)^2*(1/36))=5.8333$$

Another formula for the variance of a discrete probability distribution, VAR(X), is ...

$$VAR(X)=\sum\left[X^2 * P(X)\right]-E(X)^2$$

... where VAR(X) is the variance, X are the outcomes, P(X) are the probabilities of the outcomes, and E(X) is the mean of the distribution.

For example, the variance of the distribution shown in Table 15.1 on Page 51 is 5.8333:

$$((2^2*1/36)+(3^2*2/36)+...+(12^2*1/36))-7^2=5.8333$$

Standard Deviation

Recall that the standard deviation is also a measure of dispersion. The standard deviation of a discrete probability distribution, SD(X), is simply the square root of the variance, VAR(X). For example, the standard deviation of the distribution shown in Table 15.1 on Page 51 is 2.4152:

$$SQRT(5.8333)=2.4152$$

Blank Page

Continuous Probability Distributions

Recall that a continuous distribution shows the distribution of a continuous variable. Also recall that a probability distribution is a distribution of the theoretical probabilities of the values of a variable. A continuous probability distribution is a distribution of the theoretical probabilities of a continuous variable. For example, Table 16.1 shows the theoretical probabilities of an infinite, uniformly-distributed number of fractions between 0 and 5.

Table 16.1	
X	**Probability**
0 to 1	1/5
1 to 2	1/5
2 to 3	1/5
3 to 4	1/5
4 to 5	1/5
Total	5/5

Graphs

Recall that a density curve is a graph of the frequency distribution of a continuous variable and that the area under the curve equals the frequency of the values. Continuous probability distributions are graphed as density curves and the area equals one. For example, Figure 16.1 on Page 56 shows the density curve of the theoretical probabilities of an infinite, uniformly-distributed number of fractions between 0 and 5.

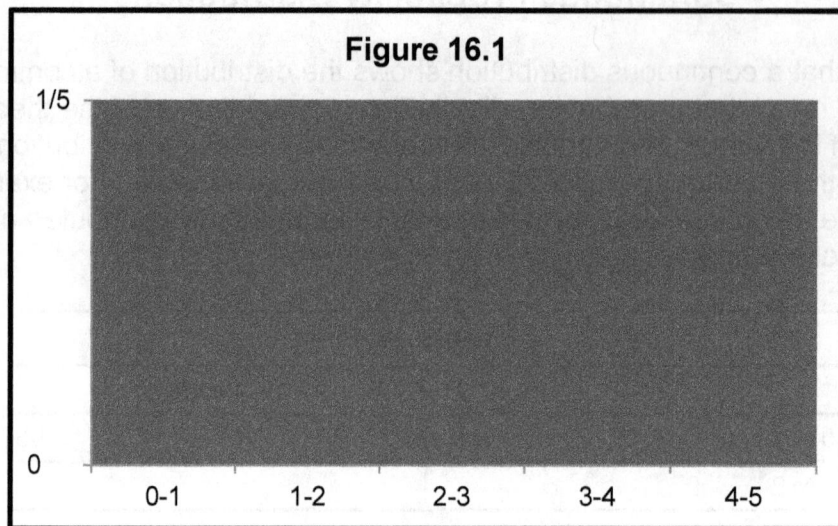

Figure 16.1

Mean

Recall that the mean is a measure of central tendency. The formula for the approximate mean of a continuous probability distribution, E(X), is ...

$$E(X)=\sum \bar{X} * P(X)$$

... where E(X) is the mean, X-bar are the middle values of the range, and P(X) are the probabilities of the range.

For example, the mean of the distribution shown in Table 16.1 on Page 55 is 2.5 and is calculated as follows:

First. Calculate the middle values:

$$(0+1)/2=0.5,(1+2)/2=1.5,...,(4+5)/2=4.5$$

Second. Calculate the mean:

$$(0.5*1/5)+(1.5*1/5)+...+(4.5*1/5)=2.5$$

Variance

Recall that the variance is a measure of dispersion. The formula for the variance of a continuous probability distribution, VAR(X), is ...

$$VAR(X)=\sum \left[\bar{X}-E(X) \right]^{2} *P(X)$$

... where VAR(X) is the variance, X-bar are the middle values of the range, E(X) is the approximate mean of the distribution, and P(X) are the probabilities of the range.

For example, the variance of the distribution shown in Table 16.1 on Page 55 is 2:

$$((0.5-2.5)^{2}*(1/5))+((1.5-2.5)^{2})*(1/5))+...+((4.5-2.5)^{2}*(1/5))=2$$

Another formula for the variance of a continuous probability distribution, VAR(X), is ...

$$VAR(X)=\sum \left[X^{2} *P(X) \right]-E(X)^{2}$$

... where VAR(X) is the variance, X are the outcomes, P(X) are the probabilities of the outcomes, and E(X) is the mean of the distribution.

For example, the variance of the distribution shown in Table 16.1 on Page 55 is 2:

$$((0.5^{2}*1/5)+(1.5^{2}*1/5)+...+(4.5^{2}*1/5))-2.5^{2}=2$$

Standard Deviation

Recall that the standard deviation is also a measure of dispersion. The standard deviation of a continuous probability distribution, SD(X), is simply the square root of the variance, VAR(X)). For example, the standard deviation of the distribution shown in Table 16.1 on Page 55 is 1.4152:

$$SQRT(2)=1.4142$$

Blank Page

Binomial Distributions

In the real world, the outcome of many events is dichotomous, that is, is divided in two. For example, a baby is born either in May or not in May. A binomial experiment, also called a Bernoulli experiment, is an experiment in which there are only two mutually exclusive outcomes that are labeled "success" or "failure". For example, the outcome of a die roll is either a 1 or not a 1. In Bernoulli experiments, p is the probability of a "success" in one trial and q (which is p-1) is the probability of a "failure" in one trial. Furthermore, in Bernoulli experiments, the probabilities of the outcomes do not change, each trial is independent, and there are repeated trials.

Binomial Distribution

The binomial distribution is a discrete probability distribution and is a distribution of the theoretical probabilities of the number of "successes", r, in a number of repeated trials, n, in a Bernoulli experiment.

Example

Table 17.1 and Figure 17.1 on Page 50 show the theoretical probabilities of the number of not 1s —that is, the theoretical probabilities of the number of "successes"— in four die rolls, in which the probability of "success" is 5 out of 6, or 5⁄6, or 83-1/3%.

Table 17.1	
Number of Not 1s	**Probability**
0	0.0008
1	0.0154
2	0.1158
3	0.3858
4	0.4822
Total	1.0000

Figure 17.1

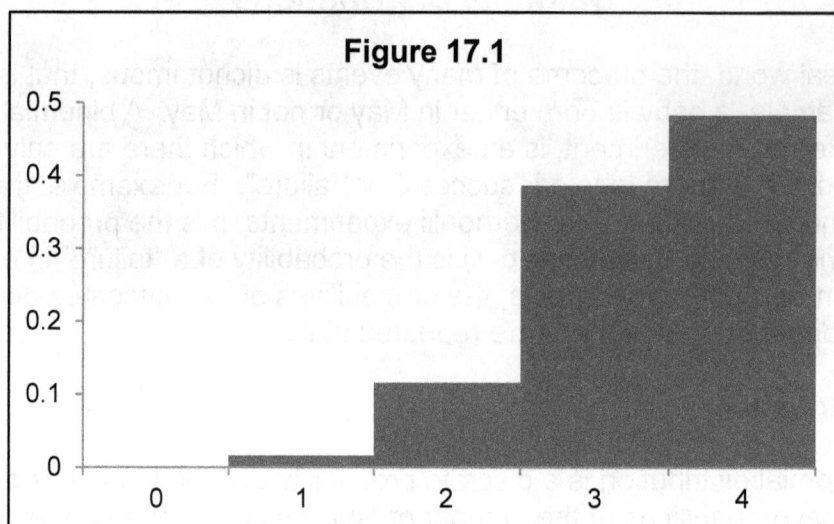

Looking at Table 17.1 on Page 59 note that the probability of getting zero not 1s —that is, no "successes"— in four die rolls is 0.0008:

$$1/6*1/6*1/6*1/6=0.0008$$

And, the probability of getting one not 1 —that is, one "success"— is 0.0154:

$$(5/6*1/6*1/6*1/6)+(1/6*5/6*1/6*1/6)+(1/6*1/6*5/6*1/6)+(1/6*1/6*1/6*5/6)=0.0154$$

In practice, as shown below, probabilities are calculated using the binomial coefficient and the binomial formula.

Binomial Coefficient

The formula for the binomial coefficient, C(n,r), is …

$$C(n,r)=\frac{n!}{r!*(n-r)!}$$

… where C(n,r) is the binomial coefficient, n is the number of trials, and r is the number of "successes".

For example, the binomial coefficient of three not 1s —that is, three "successes"— in four die rolls is 4:

$$4!/(3! * (4-3)!) = 4$$

Binomial Formula

The formula for the binomial probability, P(r), is ...

$$P(r) = C(n,r) * p^r * (1-p)^{n-r}$$

... where P(r) is the binomial probability, C(n,r) is the binomial coefficient, n is the number of trials, r is the number of "successes", and p is the probability of a "success".

For example, the probability of three not 1s —that is, three "successes"— in four die rolls, in which the probability of "success" is 5/6 is 38.58%:

$$(4!/(3! * (4-3)!)) * (5/6)^3 * (1/6)^{4-3} = 0.3858$$

Shape

As Table 17.2 shows, the shape of a binomial distribution is contingent on the probability of a "success".

<table>
<tr><td colspan="2" align="center">**Table 17.2**</td></tr>
<tr><td align="center">**Shape**</td><td align="center">**Probability**</td></tr>
<tr><td>Positively skewed</td><td align="right"><0.50</td></tr>
<tr><td>Symmetrical</td><td align="right">=0.50</td></tr>
<tr><td>Negatively skewed</td><td align="right">>0.50</td></tr>
</table>

Mean

The formula for the mean of the binomial distribution, E(X), is ...

$$E(X) = np$$

... where E(X) is the mean, n is the number of trials, and p is the probability of a "success".

For example, the mean of a binomial distribution with four trials and with a probability of a "success" of 5/6 is 3-1/3:

$$4*5/6=3-1/3$$

Variance

The formula for the variance of the binomial distribution, VAR(X), is ...

$$VAR(X)=n*p*(1-p)$$

... where VAR(X) is the variance, n is the number of trials, and p is the probability of a "success".

For example, the variance of a binomial distribution with four trials and with a probability of a "success" of 5/6 is 0.5556:

$$4*5/6*(1-5/6)=0.5556$$

Standard Deviation

The standard deviation of a binomial distribution, SD(X), is simply the square root of the variance, VAR(X). For example, the standard deviation of a binomial distribution with four trials and with a probability of a "success" of 5/6 is 0.7454:

$$SQRT(0.5556)=0.7454$$

Cumulative Probability

The cumulative probability is simply the sum of the individual probabilities. For example, the probability of three or four not 1s —that is, three or four "successes"— in four die rolls is 86.80% (See Table 17.1 on Page 59):

$$38.58+48.22=86.80$$

Normal Distributions

Recall that a density curve is a graph of the frequency distribution of a continuous variable and that the area under the curve equals the frequency of the values. In the real world, the density curves of many variables, for example, the height of adult males in the United States, show that the values are "bell-shaped" (See Figure 18.1). That is, the values gather around the middle.

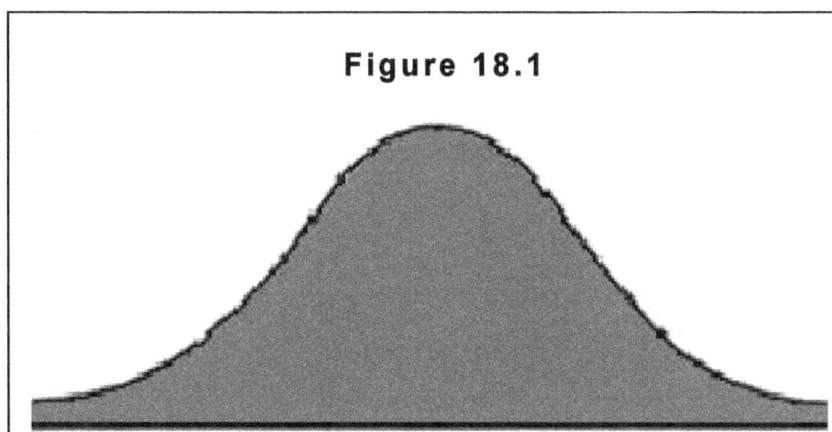

Figure 18.1

Normal Distribution

Also recall that a continuous probability distribution is a distribution of the theoretical probabilities of a continuous variable. The normal distribution is a continuous probability distribution whose features parallel the features of real-world, "bell-shaped" distributions.

Shape

Finally recall that continuous probability distributions are graphed as density curves and the area equals one. Normal distributions are graphed as density curves and the area equals one. The density curves of normal distributions are symmetric, mesokurtic, and unimodal.

Central Tendency

The mean, the median, and the mode of normal distributions are equal.

Standard Deviation

As Table 18.1 on Page 64 and Figure 18.2 on Page 64 show, about 68% of the values of normal distributions are within one standard deviation of the mean. About 95.5% of the values are within two standard deviations of the mean. And, about 99.5% of the values are within three standard deviations of the mean.

Table 18.1		
Standard Deviation	**Probability (~%)**	**Cumulative Left-Tail Probability (~%)**
-∞ to -3	0.25	0.25
-3 to -2	2.00	2.25
-2 to -1	13.75	16.00
-1 to 0	34.00	50.00
0 to +1	34.00	84.00
+1 to +2	13.75	97.75
+2 to +3	2.00	99.75
+3 to +∞	0.25	100.00

Figure 18.2

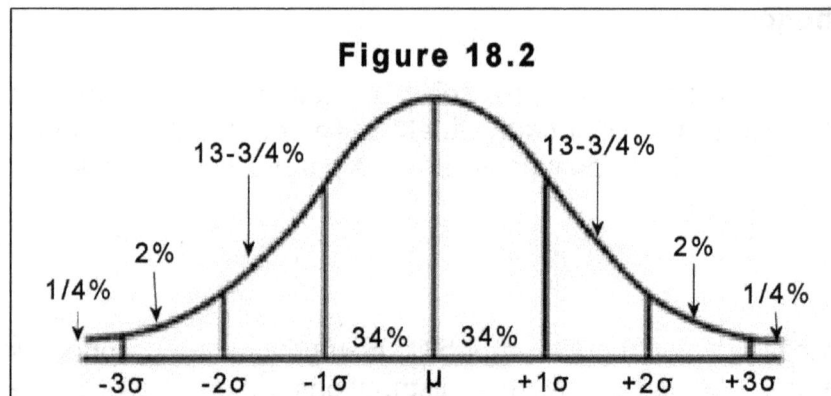

Example

The height of adult males in the United States is approximately normally distributed with a mean of 5'10" and a standard deviation of 3". That means that about 68% are between 5'7" (which is 5'10" - 3") and 6'1" (which is 5'10" + 3"). And, it also means that about 16% are 6'1" (which is 5'10" + 3") or taller.

Z Distribution

The z distribution, also called the standard normal distribution, and the normal distribution are closely related. In fact, the z distribution is a standardized normal distribution.

Z Scores

A z score, also called a z statistic or a standard score, is a measure of position that shows how many standard deviations a population value of a normally distributed population is away from the population mean. That is, a z score is the ratio of the deviation of a population value from the population mean divided by the population standard deviation ...

$$z = \frac{(X-\mu)}{\sigma}$$

... where z is the z score, X is a population value, μ is the population mean, and σ is the population standard deviation.

Z Distribution

Recall that a continuous probability distribution is a distribution of the theoretical probabilities of a continuous variable. The z distribution is a continuous probability distribution of z scores.

Shape

Also recall that continuous probability distributions are graphed as density curves and the area equals one. The z distribution is graphed as a density curve and the area equals one. The density curve of a z distribution is symmetric, mesokurtic, and unimodal.

Mean

The mean of the z distribution is equal to 0.

Standard Deviation

The standard deviation of the z distribution is equal to 1.

Probabilities

As Table 19.1 and Figure 19.1 show, about 68% of the z scores are between -1 and +1. About 95.5% of the z scores are between -2 and +2. And, about 99.5% of the z scores are between -3 and +3.

Table 19.1		
Z Scores	Probability (~%)	Cumulative Left-Tail Probability (~%)
-∞ to -3	0.25	0.25
-3 to -2	2.00	2.25
-2 to -1	13.75	16.00
-1 to 0	34.00	50.00
0 to +1	34.00	84.00
+1 to +2	13.75	97.75
+2 to +3	2.00	99.75
+3 to +∞	0.25	100.00

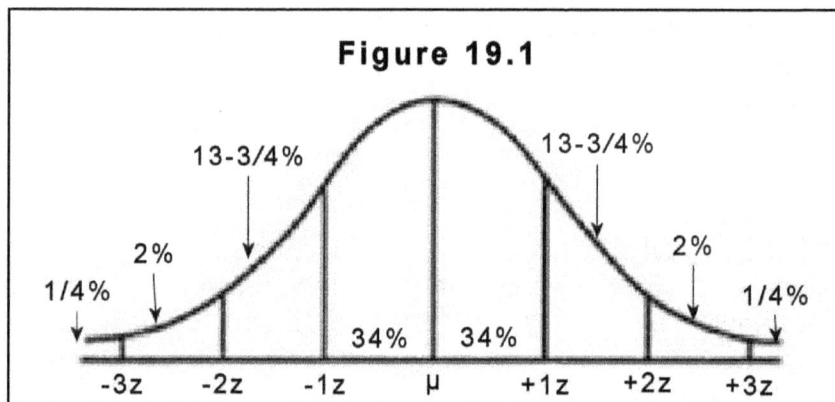

Figure 19.1

Calculation

The formula for calculating the z score of a population value of a normally distributed population is ...

$$z=\frac{(X-\mu)}{\sigma}$$

… where z is the z score, X is the population value, μ is the population mean, and σ is the population standard deviation.

Note. The formula for calculating the z score for proportions is presented in Chapter 34.

Example

The height of adult males in the United States is approximately normally distributed with a mean of 5'10" and a standard deviation of 3". The z score of someone who is 6'1" is 1:

$$(6'1"-5'10")/3"=1$$

Z Tables

Z tables show the z scores and the cumulative left-tail probabilities between -∞ and the z scores. That is, z tables show the cumulative probabilities to the left of z scores. For example, if the z score is -1.00, then, about 16% of the z scores are between -∞ and -1.00 (See Table 19.2 and Figure 19.2).

Table 19.2		
z Score		**Cumulative Left-Tail Probability**
-1.0	.00	0.15866

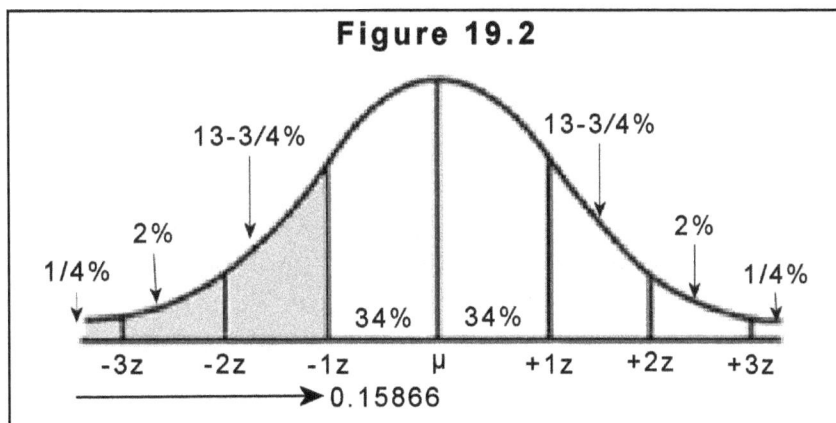

Figure 19.2

And, if the z score is +1.75, then, about 96% of the z scores are between -∞ and +1.75 (See Table 19.3 on Page 68 and Figure 19.3 on Page 68).

Table 19.3	
z Score	**Cumulative Left-Tail Probability**
+1.7 .05	0.95994

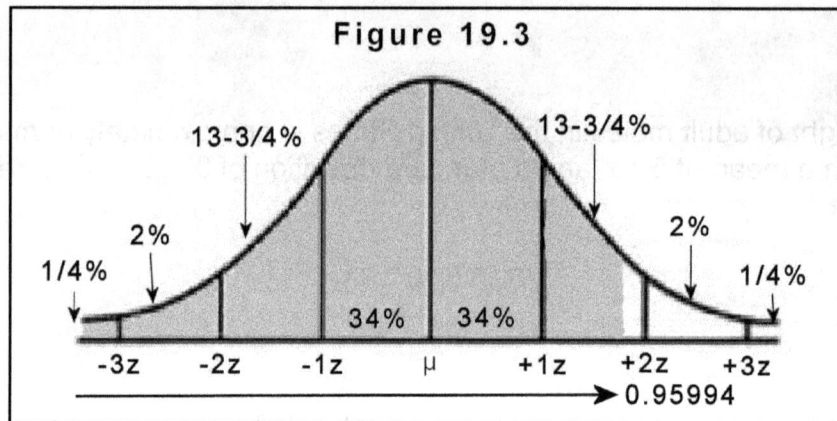
Figure 19.3

Critical Values, Significance Level, and Critical Region

In inferential statistics, z tables (See Table 19.4) are used to find the critical values (CV) of z. CVs are the z scores that correspond to a level of significance, called the alpha level, α, which delineates the critical region.

Table 19.4		
z Score		**Cumulative Left-Tail Probability**
-1.9	.60	0.02750
-1.6	.45	0.05000
+1.6	.45	0.95000
+1.9	.60	0.97500

For example, Table 19.4 and Figure 19.4 on Page 69 show the right critical value of z=+1.645 and the corresponding critical region of α=5%. The critical region shows the probability, which is 5%, that the z scores are more than +1.645. Note that z tables show the cumulative probability to the left of the critical value.

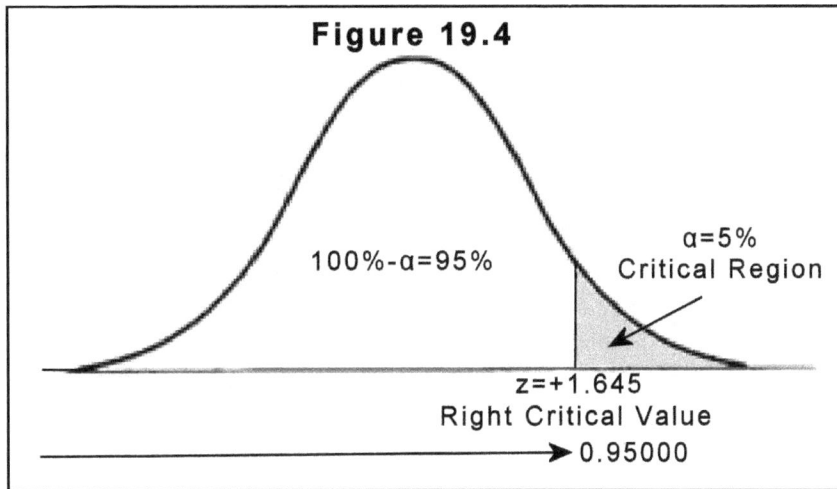
Figure 19.4

And, Table 19.4 on Page 68 and Figure 19.5 show the left critical value of z=-1.645 and the corresponding critical region of α=5%. The critical region shows the probability, which is 5%, that the z scores are less than -1.645. Note that z tables show the cumulative probability to the left of the critical value.

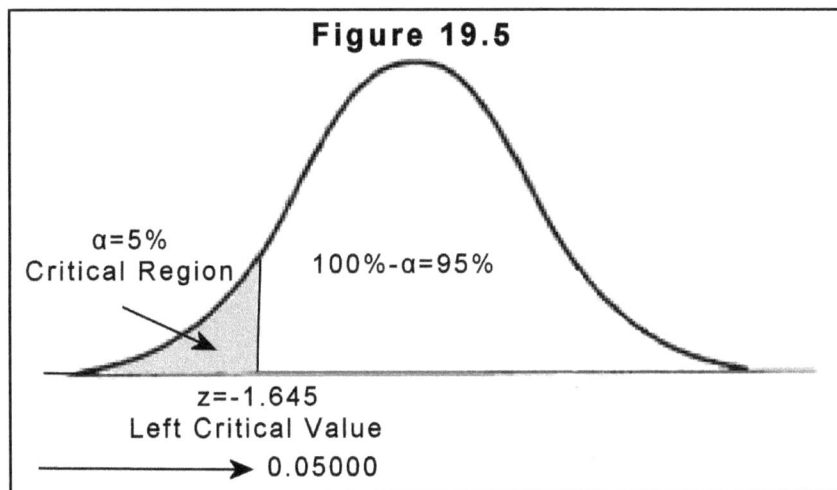
Figure 19.5

Finally, Table 19.4 on Page 68 and Figure 19.6 on Page 70 show the left and the right critical values of z=-1.96 and z=+1.96 and the corresponding critical region of α=5%. Note that the critical region is split in two. The critical regions shows the probability, which is 5%, that the z scores are less than -1.96 or more than +1.96. Note that z tables show the cumulative probability to the left of the critical value.

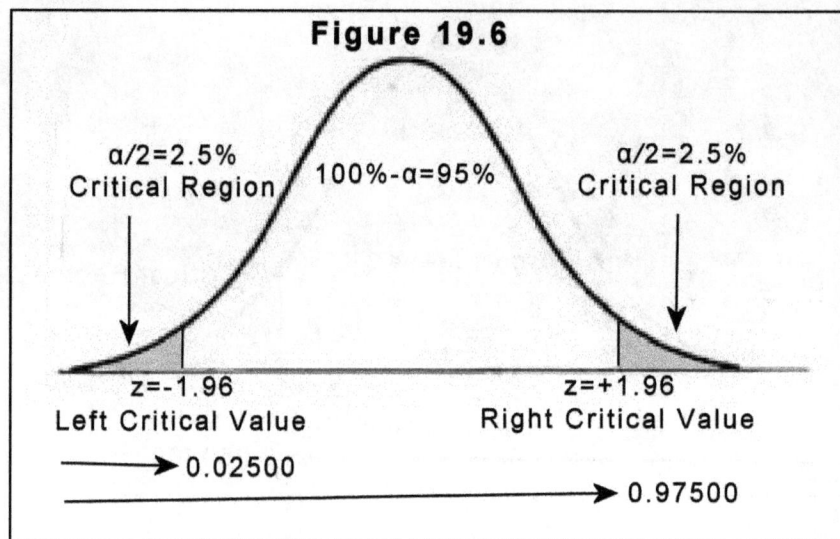

Figure 19.6

α/2=2.5%
Critical Region

100%-α=95%

α/2=2.5%
Critical Region

z=-1.96
Left Critical Value

z=+1.96
Right Critical Value

→ 0.02500

→ 0.97500

T Distributions

As Figure 20.1 shows, t distributions and the z distribution are closely related. Note that t distributions are spread out more than the z distribution.

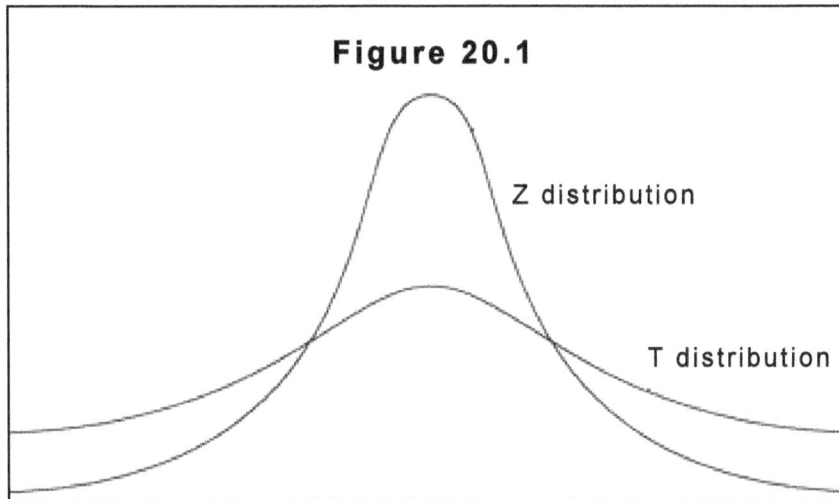

Figure 20.1

Z distribution

T distribution

The main difference between the t distributions and the z distribution is that t distributions are contingent on the sample size, called "degrees of freedom". That is, there is one z distribution but there are many t distributions (See Figure 20.2). As the sample size increases, t distributions approximate the z distribution.

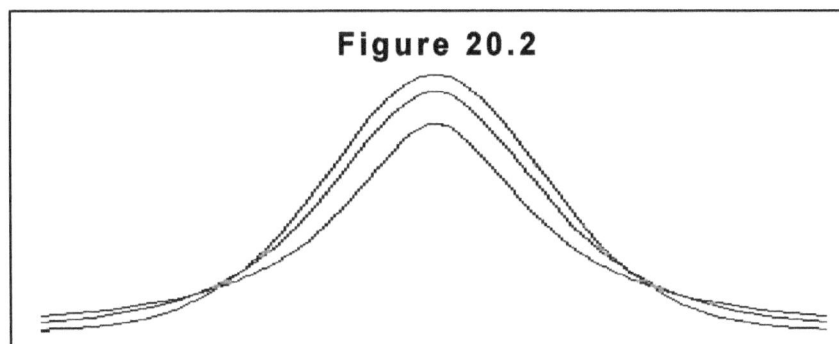

Figure 20.2

Standard Error

Recall that the standard error measures the variability —that is, the spread— of statistics of samples. The standard error of the sample mean. $Se_{x\text{-bar}}$, is a measure of the variability of the sample mean and is equal to …

$$SE_{\bar{x}} = \frac{s}{\sqrt{n}}$$

… where $SE_{x\text{-bar}}$ is the standard error of the sample mean, s is the standard deviation of the sample, and n is the sample size.

T Scores

A t score, also called a t statistic, is a measure of position that shows how many standard errors of the sample mean the sample mean of a population, which might or might not be normally distributez, is away from the population mean. That is, a t score is the ratio of the deviation of a sample mean from the population mean divided by the standard error of the sample mean …

$$t = \frac{\bar{x} - \mu}{\frac{s}{\sqrt{n}}}$$

… where t is the t score, x-bar is the sample mean, μ is the population mean, s is the standard deviation of the sample, and n is the sample size. As shown before, the standard error of the sample mean, $SE_{x\text{-bar}}$, equals s divided by the square root of n.

T Distributions

Recall that a continuous probability distribution is a distribution of the theoretical probabilities of a continuous variable. The t distribution is a continuous probability distribution of t scores.

Degrees of Freedom

The degrees of freedom (DF) of t distributions are equal to the sample size minus one. That is, DF=n-1.

Shape

Also recall that continuous probability distributions are graphed as density curves and the area equals one. The t distributions are graphed as density curves and the area equals one. The density curve of a t distribution is symmetric, platykurtic, and unimodal. The degree of platykurtosis of t distributions is contingent on the degrees of freedom –that is, the sample size.

Mean

The mean of t distributions is equal to 0.

Standard Deviation

The standard deviation of t distributions is greater than one. As the degrees of freedom increase —that is, as the sample size increases— the standard deviation of t distributions approximates 1.

Probabilities

As Table 20.1 and Figure 20.3 on Page 74 show, for DF=15 (which is n=16), about 67% of the t scores are between -1 and +1. About 94% of the t scores are between -2 and +2. And, about 99% of the t scores are between -3 and +3.

As the degrees of freedom increase —that is, as the sample size increases— the probabilities of the t distributions approximate the probabilities of the z distribution.

Table 20.1		
DF=15		
T Scores	Probability (~%)	Cumulative Left-Tail Probability (~%)
-∞ to -3	0.45	0.45
-3 to -2	2.75	3.20
-2 to -1	13.46	16.66
-1 to 0	33.34	50.00
0 to +1	33.34	83.34
+1 to +2	13.46	96.80
+2 to +3	2.75	99.55
+3 to +∞	0.45	100.00

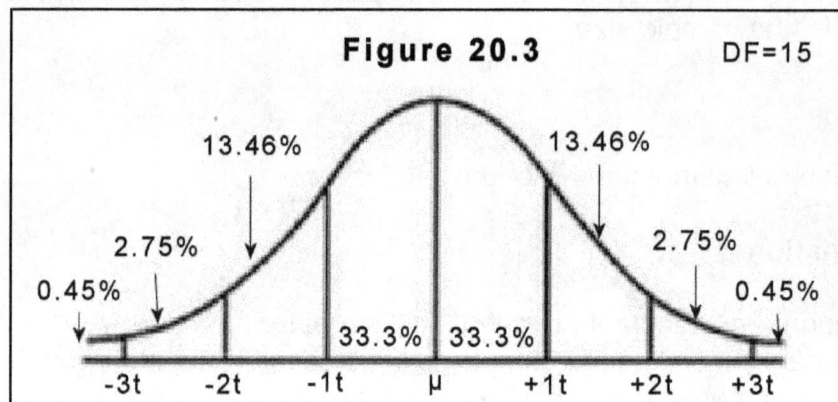

Figure 20.3   DF=15

13.46%   13.46%

2.75%   2.75%

0.45%   0.45%

33.3% | 33.3%

-3t   -2t   -1t   μ   +1t   +2t   +3t

Calculation

The formula for calculating the t score of a sample mean of a population, which may or may not be normally distributed, is ...

$$t=\frac{\bar{x}-\mu}{\frac{s}{\sqrt{n}}}$$

... where t is the t score, x-bar is the sample mean, μ is the population mean, s is the standard deviation of the sample, and n is the sample size. As shown before, the standard error of the sample mean, $SE_{x\text{-}bar}$, equals s divided by the square root of n.

Example

The height of adult males in the United States is approximately normally distributed with a mean of 5'10". For a sample of ten with a mean of 5'11" and a standard deviation of 3.1", the t score is 1.02:

(5'11"-5'10")÷[3.1"÷SQRT(10-1)]=1.02

T Tables

T tables show the t scores for selected degrees of freedom and selected probabilities. Caveat! For one tail, t tables only show right-tail probabilities. But, t distributions are symmetrical. Therefore, left-tail probabilities and right-tail probabilities are equal. For example, as Table 20.2 on Page 75 and Figure 20.4 on Page 75 show, for DF=15, if the t score is +2.131, then, about 2.5% of the t scores are more than

+2.131.

Table 20.2			
Degrees of Freedom	**Probability (One Tail)**		
	0.1000	0.0500	0.0250
	1.341	1.753	2.131
15	**Probability (Two Tail)**		
	0.1000 (0.0500+0.0500)	0.0500 (0.0250+0.0250)	0.0250 (0.0125+0.0125)
	1.753	2.131	2.490

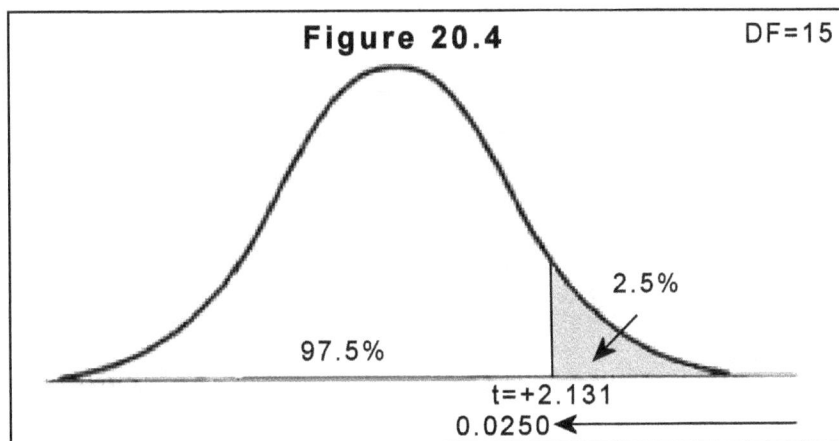

Figure 20.4   DF=15

97.5%   2.5%

t=+2.131
0.0250 ◄─────────────

And, as Table 20.2 and Figure 20.5 on Page 76 show, for DF=15, if the t score is -2.131, then, about 2.5% of the t scores are less than -2.131.

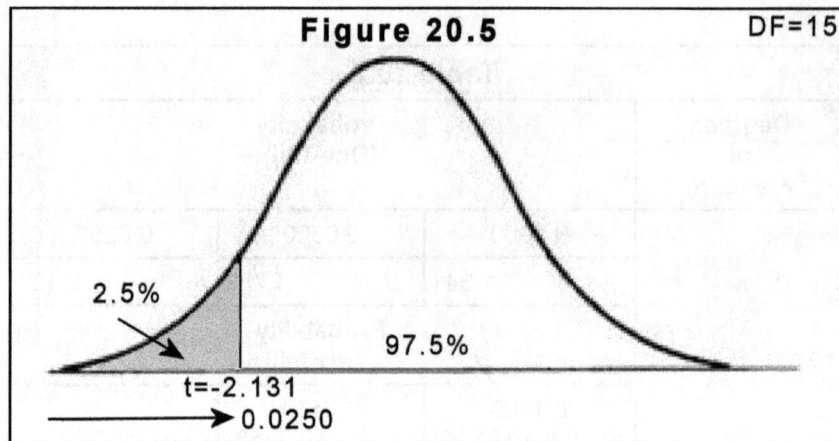

Figure 20.5   DF=15

2.5%

97.5%

t=-2.131

0.0250

Finally, as Table 20.2 on Page 75 and Figure 20.6 show, for DF=15, about 2.5% of the t scores are less than -2.131 and about 2.5% of the t scores are more than +2.131.

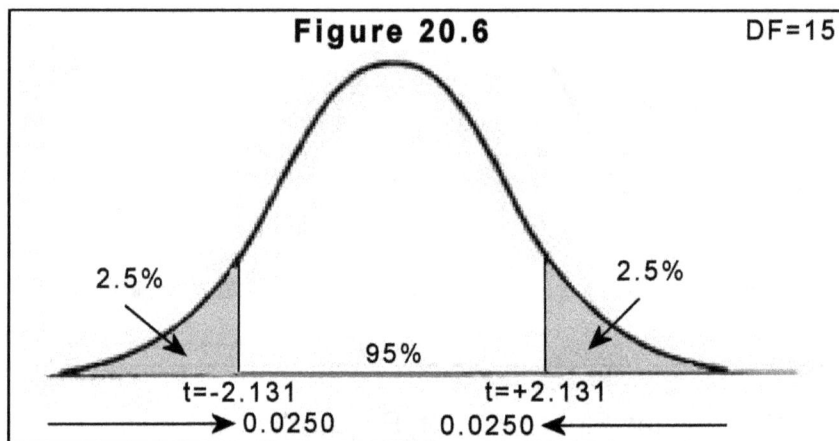

Figure 20.6   DF=15

2.5%   2.5%

95%

t=-2.131   t=+2.131

0.0250   0.0250

Critical Values, Significance Level, and Critical Region

In inferential statistics, t tables (See Table 20.2 on Page 75) are used to find the critical values (CV) of t. CVs are the t scores that correspond to a level of significance, called the alpha level, α, which delineates the critical region.

For example, Table 20.2 on Page 75 and Figure 20.7 on Page 77 show, for DF=15, the right critical value of t=+1.753 and the corresponding critical region of α=5%. The critical region shows the probability, which is 5%, that the t scores are more than +1.753. Note that t tables show the probability to the right of the critical value.

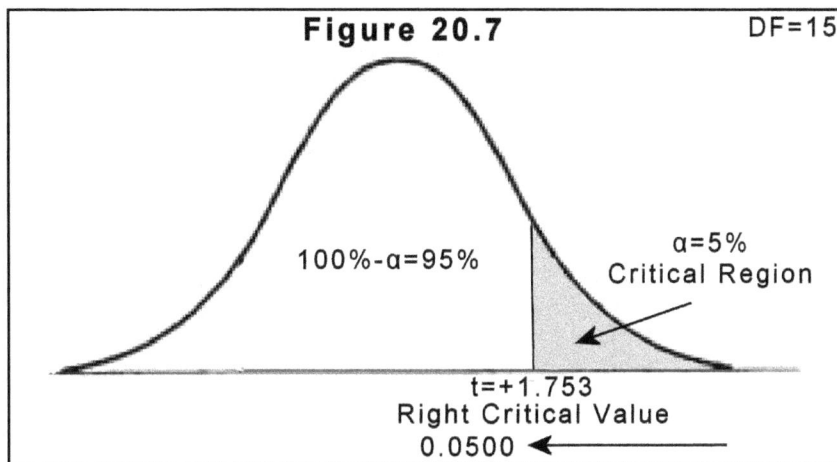

Figure 20.7, DF=15

And, Table 20.2 on Page 75 and Figure 20.8 show, for DF=15, the left critical value of t=-1.753 and the corresponding critical region, α=5%. The critical region shows the probability, which is 5%, that the t scores are less than -1.753. Note that t tables show the probability to the left of the critical value.

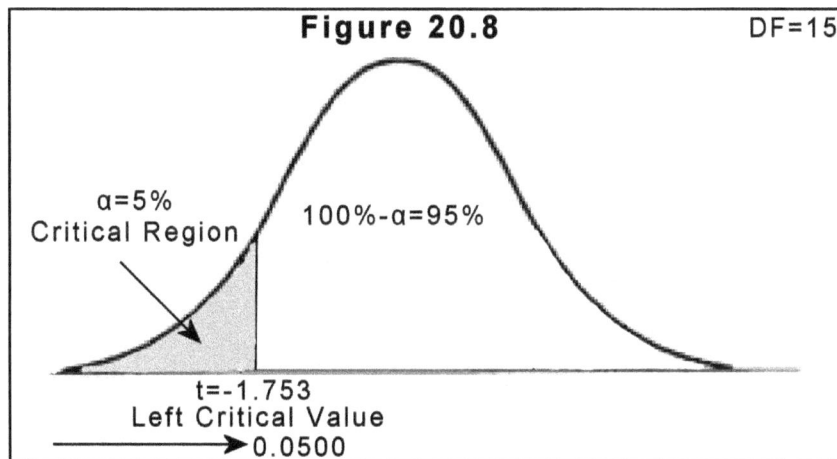

Figure 20.8, DF=15

Finally, Table 20.2 on Page 75 and Figure 20.9 on Page 78 show, for DF=15, the left and the critical values of t=-2.131 and t=+2.131 and the corresponding critical region of α=5%. Note that the critical region is split in two. The critical regions show the probabilities, which is 5%, that the t scores are less than -2.131 or more than +2.131. Note that t tables show the probabilities to the left and to the right of the critical values.

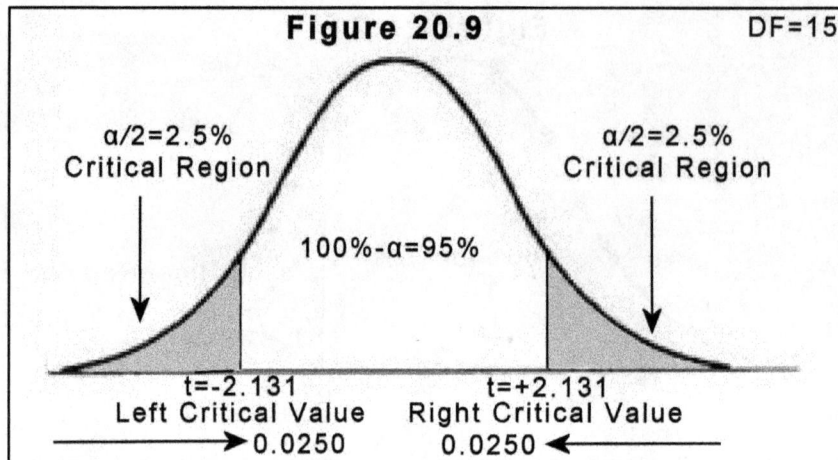

Figure 20.9   DF=15

$\alpha/2=2.5\%$
Critical Region

$\alpha/2=2.5\%$
Critical Region

$100\%-\alpha=95\%$

t=-2.131
Left Critical Value

t=+2.131
Right Critical Value

0.0250   0.0250

Chi-Square Distributions

Chi-square distributions and the z distribution are closely related. In fact, z scores are the building blocks of chi-square statistics.

Say that one draws an infinite number of samples of size n=1 from a z distribution and, then, squares the z scores. The squared z scores, called chi-square statistics, X^2, make up a chi-square distribution with one degree of freedom. Now, say that one draws an infinite number of samples of size n=2 from the z distribution, again squares the z scores, and, then, adds the squared z scores. The sums of the squared z scores make up a chi-square distribution with two degrees of freedom. Chi-square statistics are never negative and the degrees of freedom of the chi-square distribution equal the sample size —that is, the number of squared z scores being added.

Chi-Square Statistics

Chi-square statistics, X^2, are sums of squared z scores and are measures of position. As stated before, chi-square statistics are never negative and the degrees of freedom of the chi-square distribution equal the sample size —that is, the number of squared z scores being added.

Chi-Square Distributions

Recall that a continuous probability distribution is a distribution of the theoretical probabilities of a continuous variable. The chi-square distributions are continuous probability distributions of chi-square statistics.

Degrees of Freedom

The degrees of freedom of chi-square distributions are equal to the sample size —that is, the number of squared z- scores being summed (See Figure 21.1 on Page 80).

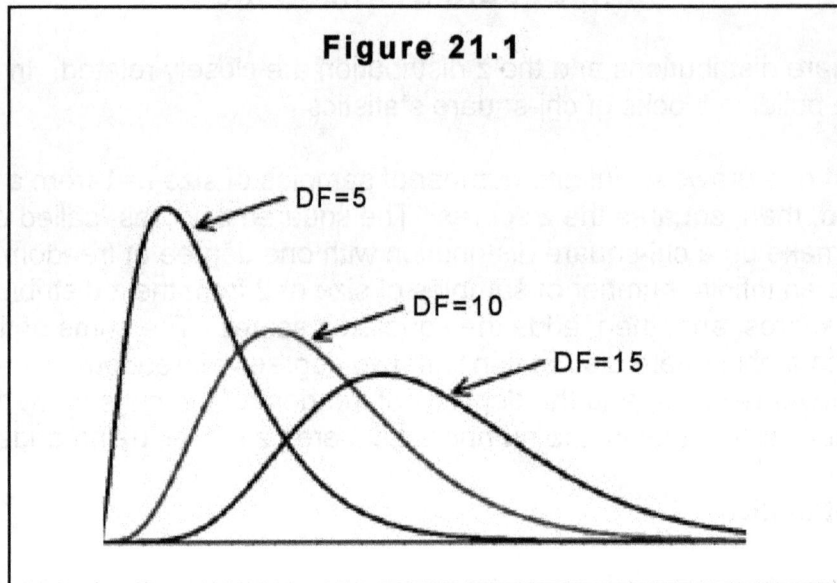

Figure 21.1

DF=5
DF=10
DF=15

Shape

Also recall that continuous probability distributions are graphed as density curves and the area equals one. The chi-square distributions are graphed as density curves and the area equals one. The density curve of a chi-square distribution is asymmetric, right-skewed, and unimodal. As the degrees of freedom increases —that is, as the sample size increases— the chi-square distributions approach a normal distribution.

Mean

The mean of chi-square distributions is equal to the degrees of freedom.

Standard Deviation

The standard deviation of chi-square distribution is equal to the square root of twice the degrees of freedom.

Probabilities

As Table 21.1 on Page 81 and Figure 21.2 on Page 81 show, for DF=5, about 5% of the chi-square statistics are more than 11.07.

Table 21.1	
DF=5	
Chi-square Statistics	**Cumulative Right-Tail Probability (~%)**
∞ to 0	100.00
∞ to 0.83	97.50
∞ to 1.15	95.00
∞ to 1.61	90.00
∞ to 9.24	10.00
∞ to 11.07	5.00
∞ to 12.83	2.50

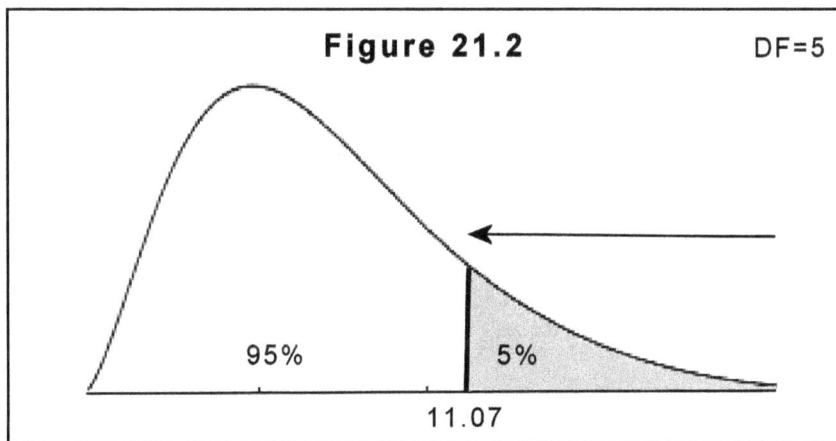

Figure 21.2   DF=5

95%   5%

11.07

Chi-Square Tables

Chi-square tables show the chi-square statistics for selected degrees of freedom and selected right-tail probabilities —that is alpha, α. For example, as Table 21.2 on Page 82 and Figure 21.2 show, for DF=5, the chi-square statistic is 11.070 and about 5% of the chi-square statistics are more than 11.070 .

Table 21.2				
Degrees of Freedom	**Probability (Right-Tail)**			
	0.975	0.950	0.050	0.025
5	0.831	1.145	11.070	12.833

Critical Values, Significance Level, and Critical Region

In inferential statistics, chi-square tables (See Table 21.3) are used to find the critical values (CV) of chi-square. CVs are the chi-square statistics that correspond to a level of significance, called the alpha level, α, which delineates the critical region.

Table 21.3				
Degrees of Freedom	**Probability (Right-Tail)**			
	0.975	0.950	0.050	0.025
15	6.262	7.261	24.996	27.488

For example, Table 21.3 and Figure 21.3 show, for DF=15, the right critical value of X^2=24.996 and the corresponding critical region of α=5%. The critical region shows the probability, which is 5%, that the X^2 statistics are more than 24.996. Note that chi-square tables show the probability to the right of the critical value.

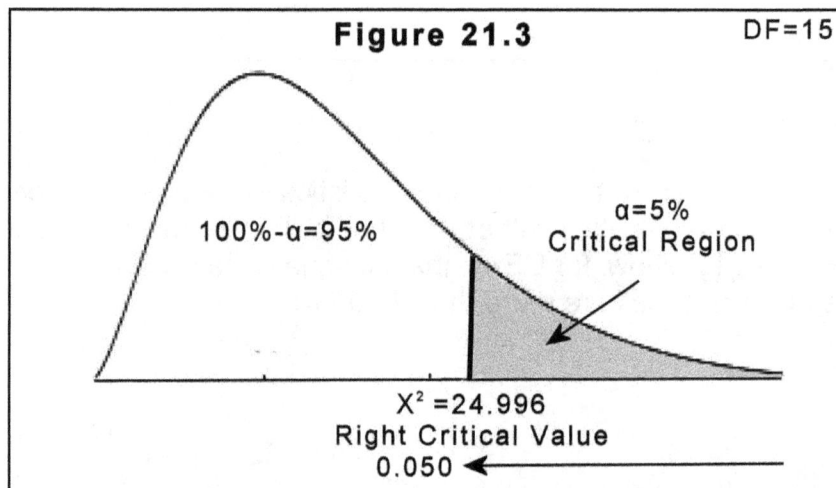

Figure 21.3   DF=15

100%-α=95%

α=5%
Critical Region

X^2 =24.996
Right Critical Value
0.050

And, Table 21.3 and Figure 21.4 on Page 83 show, for DF=15, the left critical value of X^2=7.261 and the corresponding critical region of α=5%. The critical region shows the probability, which is 5%, that the X^2 statistics are less than 7.261. Note that chi-square tables show the probability to the right of the critical value.

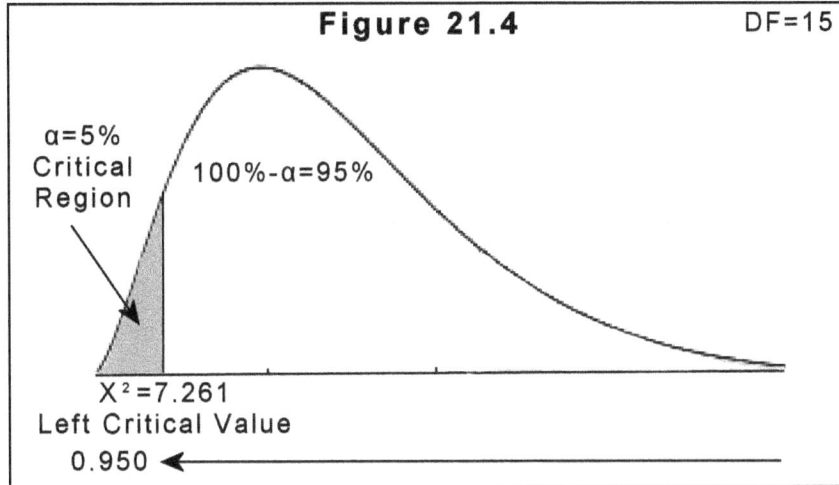

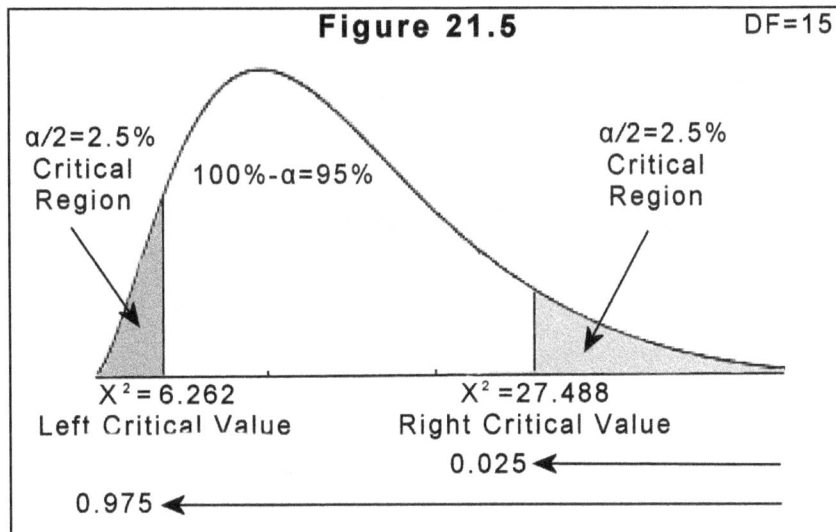

Figure 21.4   DF=15

α=5%
Critical
Region   100%-α=95%

X^2=7.261
Left Critical Value
0.950 ←

Finally, Table 21.3 on Page 82 and Figure 21.5 show for DF=15, the left and the right critical values of X^2=6.262 and X^2=27.488 and the corresponding critical region of α=5%. Note that the critical region is split in two. The critical regions show the probability, which is 5%, that the X^2 statistics are less than 6.262 or more than 27.488. Note that chi-square tables show the probability to the right of the critical value.

Figure 21.5   DF=15

α/2=2.5%
Critical
Region   100%-α=95%

α/2=2.5%
Critical
Region

X^2= 6.262
Left Critical Value

X^2 =27.488
Right Critical Value

0.025 ←

0.975 ←

Blank Page

F Distributions

F distributions and chi-square distributions are closely related. In fact, chi-square statistics are the building blocks of f statistics.

Say one takes a chi-square statistic, X^2, with DF=p and divides X^2 by p, then, one takes another X^2 with DF=q and divides X^2 by q, and, finally, one divides the preceding two outcomes. The final outcome is an f statistic with DF=p,q. That is ...

$$F_{p,q} = \frac{\dfrac{X_p^2}{p}}{\dfrac{X_q^2}{q}}$$

... where $F_{p,q}$ is the F statistic with DF=p,q, X^2 are the chi-square statistics with p and q degrees of freedom, respectively, and p and q are the degrees of freedom.

F Statistics

An f statistic is the ratio of a chi-square statistic over the chi-square statistic's degrees of freedom divided by another chi-square statistic over the chi-square statistic's degrees of freedom.

F Distributions

Recall that a continuous probability distribution is a distribution of the theoretical probabilities of a continuous variable. The f distributions are continuous probability distributions of f statistics.

Degrees of Freedom

The two degrees of freedom of f distributions are the numerator's degree of freedom and the denominator's degree of freedom See Figure 22.1 on Page 86).

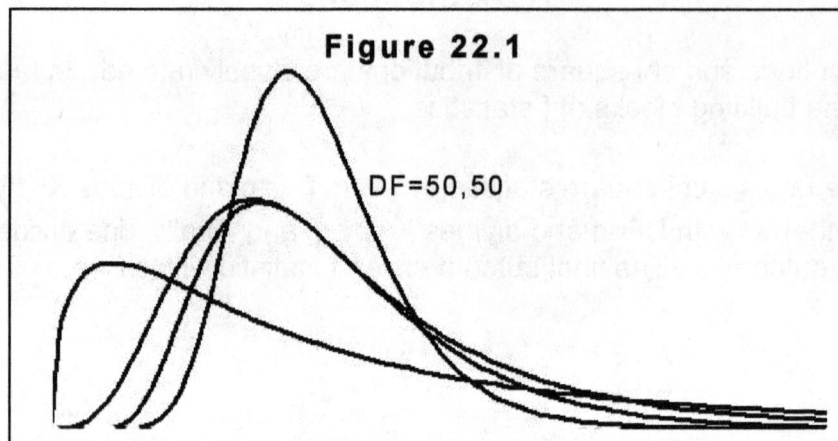

Figure 22.1

DF=50,50

Shape

Also recall that continuous probability distributions are graphed as density curves and the area equals one. The f distributions are graphed as density curves and the area equals one. The density curve of an f distribution is asymmetric, right-skewed, and unimodal.

Mean

The formula for the mean of f distributions is ...

$$\mu = \frac{DF_2}{DF_2 - 2}$$

... where μ is the mean and DF_2 is the degrees of freedom in the denominator, for $DF_2 > 2$.

Variance

The formula for the variance of f distributions is ...

$$\sigma^2 = \frac{2 * DF_1^2 * (DF_1 + DF_2 - 2)}{DF_2 * (DF_1 - 2)^2 * (DF_1 - 4)}$$

... where σ^2 is the variance, DF_1 is the the degrees of freedom in the numerator, and DF_2 is the the degrees of freedom in the denominator.

Probabilities

As Table 22.1 and Figure 22.2 show, for DF=50,50, about 5% of the f statistics are more than 1.60.

Table 22.1	
DF=50,50	
F Statistics	**Cumulative Right-Tail Probability (~%)**
∞ to 0	100.00
∞ to 0.57	97.50
∞ to 0.63	95.00
∞ to 0.69	90.00
∞ to 1.44	10.00
∞ to 1.60	5.00
∞ to 1.75	2.50

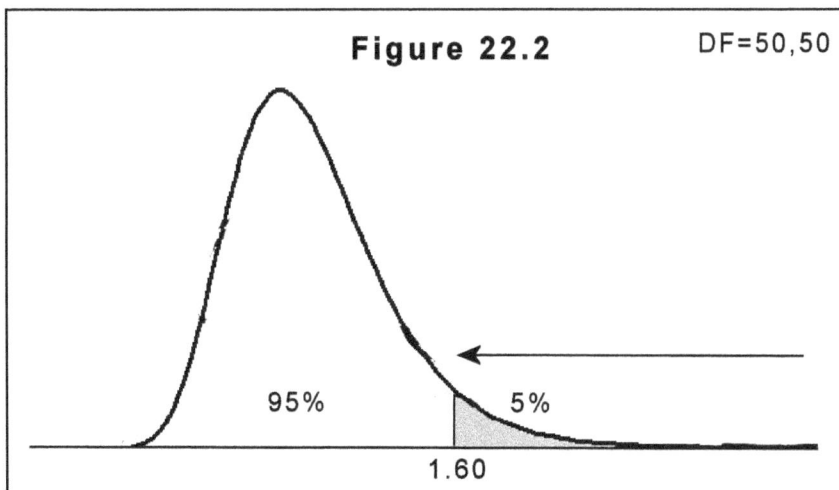

Figure 22.2   DF=50,50

95%   5%

1.60

F Tables

F tables show the f statistics for selected pairs of degrees of freedom and selected right-tail probabilities —that is alpha, α. For example, as Table 22.2 on Page 87 and Figure 22.2 on Page 87 show, for α=5%, the f statistic, for DF=50,50, is 1.60 and about 5% of the f statistics are more than 1.60.

Table 22.2			
α=5%			
	DF$_1$=1	DF$_1$=50	DF$_1$=∞
DF$_2$=1	161.45	251.77	254.32
DF$_2$=50	4.03	1.60	1.44
DF$_2$=∞	3.84	1.35	1.00

Critical Values, Significance Level, and Critical Region

In inferential statistics, f tables (See Table 22.3) are used to find the critical values (CV) of f. CVs are the f statistics that correspond to a level of significance, called the alpha level, α, which delineates the critical region.

Table 22.3		
α=5%		
	DF$_1$=5	DF$_1$=6
DF$_2$=5	5.0503	4.9503
DF$_2$=6	4.3874	4.2839

For example, Table 22.3 and Figure 22.3 on Page 89 shows, for DF=5,6, the right critical value of f=4.3874 and the corresponding critical region of α=5%. The critical region shows the probability, which is 5%, that the f statistics are more than 4.3874. Note that f tables show the probability to the right of the critical value.

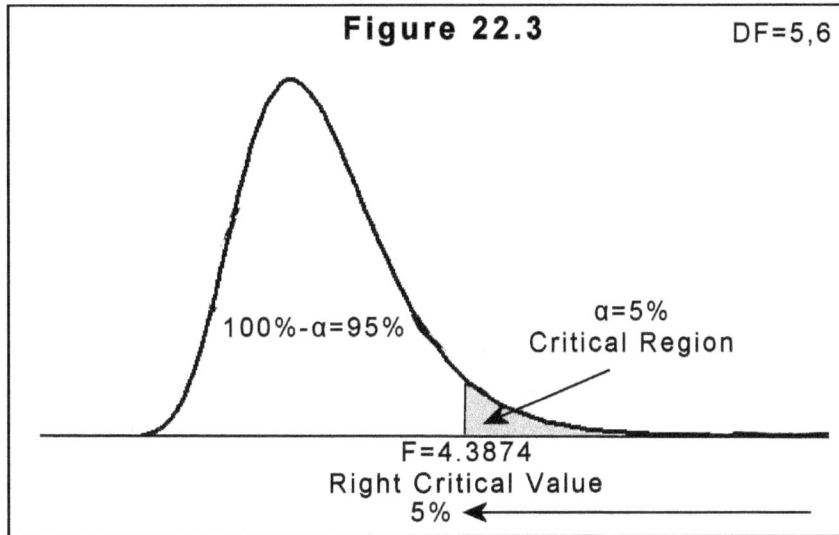

Figure 22.3   DF=5,6

100%-α=95%

α=5%
Critical Region

F=4.3874
Right Critical Value
5%

And, Figure 22.4 shows, for DF=5,6, the left critical value of f=0.2020 and the corresponding critical region of α=5%. The critical region shows the probability, which is 5%, that the f statistics are less than 0.2020. The left critical value is found as follows:

- Flip the degrees of freedom and find the f statistic for α=5% and DF=6,5, which is 4.9503 (See Table 22.3 on Page 88).
- Find the reciprocal of 4.9503, which is 1/4.9503 or 0.2020.

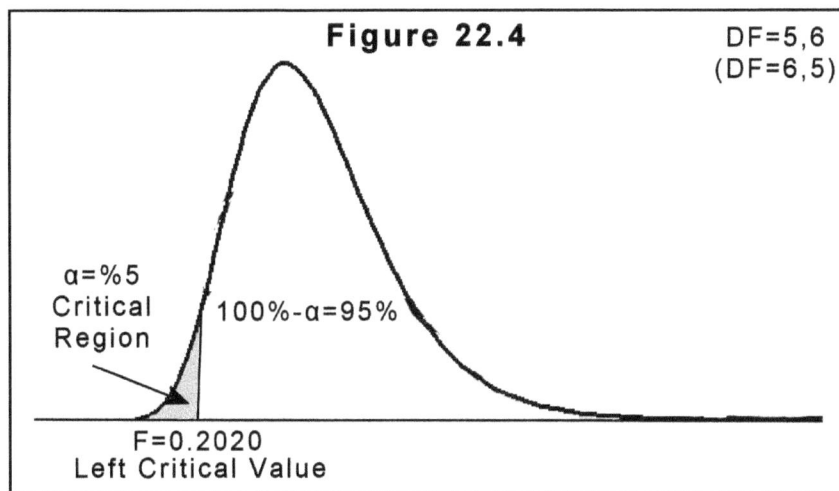

Figure 22.4   DF=5,6
                                                 (DF=6,5)

α=%5
Critical
Region

100%-α=95%

F=0.2020
Left Critical Value

Finally, Figure 22.5 on Page 90 shows for DF=5,6, the left and the right critical values of f=0.2020 and f=4.3874 and the corresponding critical region of α=10%. Note that the critical region is split in two. The critical regions show the probability, which is

10%, that the f statistics are less than 0.2020 or more than 4.3874. The right critical value is found directly from the f tables (See Table 22.3). Note that f tables show the probability to the right of the critical value. And, the left critical value is found as follows:

- Flip the degrees of freedom and find the f statistic for $\alpha/2=5\%$ and DF=6,5, which is 4.9503 (See Table 22.3 on Page 88).
- Find the reciprocal of 4.9503, which is 1/4.9503 or 0.2020.

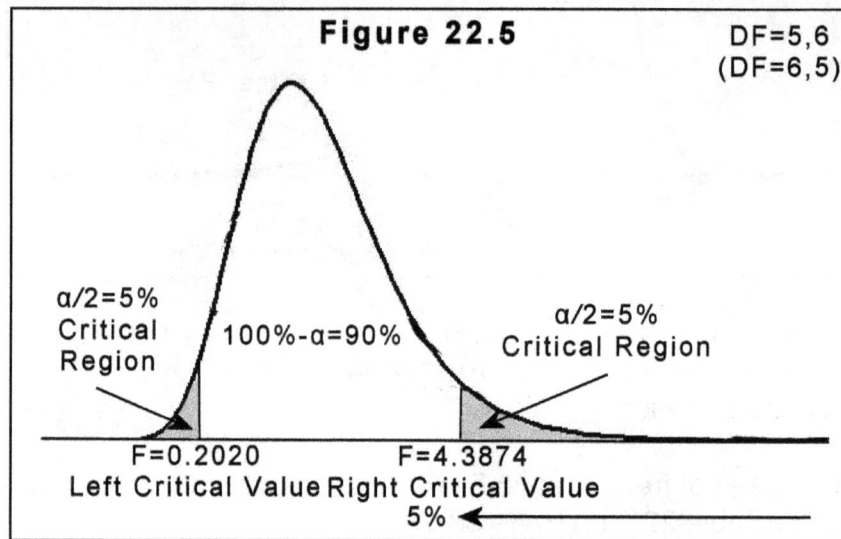

Figure 22.5

DF=5,6
(DF=6,5)

$\alpha/2=5\%$ Critical Region

100%-α=90%

$\alpha/2=5\%$ Critical Region

F=0.2020
Left Critical Value

F=4.3874
Right Critical Value

5%

In practice, only the right critical value is found (See Chapter 37).

Introduction to Inference

Recall that inferential statistics are ways of drawing inferences from data. Inferential statistics are:

- Ways of estimating the characteristics of populations based on the characteristics of samples drawn from the populations. For example, one might estimate the average weight of the United States population by weighing a portion of the population.

- Ways of comparing the characteristics of populations based on the characteristics of samples drawn from the populations. For example, one might say that the life expectancy of Hispanic American males is about 80 years and, then, compare the life expectancy of a sample of Hispanic American males to the hypothesized value of 80.

Briefly stated, the two main tasks of inferential statistics are estimation and hypothesis testing.

Estimation

Recall that parameters are values that describe the characteristics of populations whereas statistics are values that describe the characteristics of samples. Also recall that, oftentimes, the parameters of populations are unknown and the statistics of samples are used to estimate the parameters of populations. Estimation includes point estimation and interval estimation. For example, one may say that, in the U.S., the average height of all adult males is about 70 inches, that is, 5 feet, 10 inches. And, one may say that one is "confident" that, in 95% of the times, the average height is within an interval of 64.12 inches and 75.88 inches.

Hypothesis Testing

A hypothesis is a statement about a characteristic of a population —that is, the variable to be tested. For example, one might say that, in the U.S., the average household income is about $52 thousand per year. Simply stated, hypothesis tests, also called. significance tests, are ways for failing to reject —that is, accepting— or rejecting such statements.

Probability

Estimates and the results of hypothesis tests are expressed in terms of probabilities. For example, comparing a sample mean, x-bar, to a population mean, μ, using z score, Figure 23.1 shows the probability of the sample mean being way out to the right of the population mean.

Figure 23.1

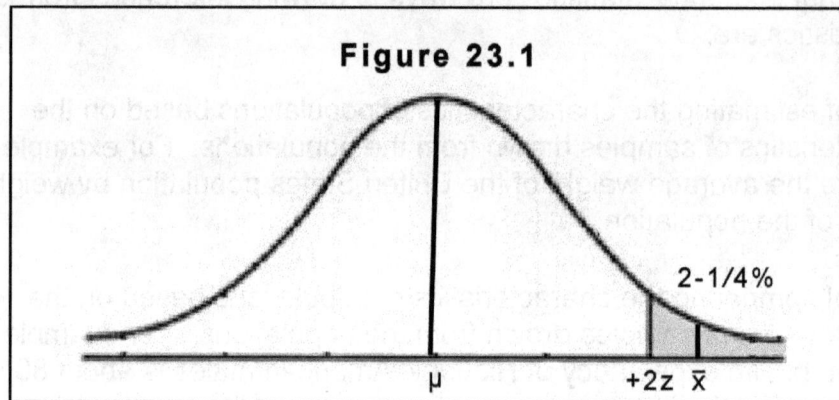

Sampling Error

Statistics of samples are estimates of population parameters. Statistics of samples deviate from population parameters by the sampling error.

Statistical Significance

Statistical significance shows the probability that results are or are not due to chance —that is, due to sampling error. That is, statistical significance is a measure of reliability.

Methodology

In inferential statistics, the methodology used for estimation and hypothesis testing is contingent on the shape of the distribution of the population, the sample size, and whether or not the standard deviation of the population is known (See Table 23.1 on Page 93).

Table 23.1			
Normal Population			
n>30		**n<30**	
σ **Known**	σ **Unknown**	σ **Known**	σ **Unknown**
z Score	z or t Score*	z Score	t Score
Not Normal Population			
n>30		**n<30**	
σ **Known**	σ **Unknown**	σ **Known or Unknown**	
z Score	t Score	Non-parametric Method	

    * According to some, but not to all, if the sample size is greater than 30, then one uses a z score rather than a t score.

Blank Page

Sampling Distributions

Recall that parameters are values that describe the characteristics of populations whereas statistics are values that describe the characteristics of samples. Intuitively, different samples of a population yield different statistics, for example, different means. This chapter applies to all statistics, for example, the median, however, for simplicity, the focus of this chapter is on means only.

Recall that a distribution is a set of the values of a variable. Say one calculates the means of either (1) selected random samples of size n from a population or (2) all possible samples of size n from a population. Either way, the distribution of the sample means is called the sampling distribution of the sample means. Thus, the mean of the sampling distribution of the sample means is the "mean of means" (See Figure 24.1).

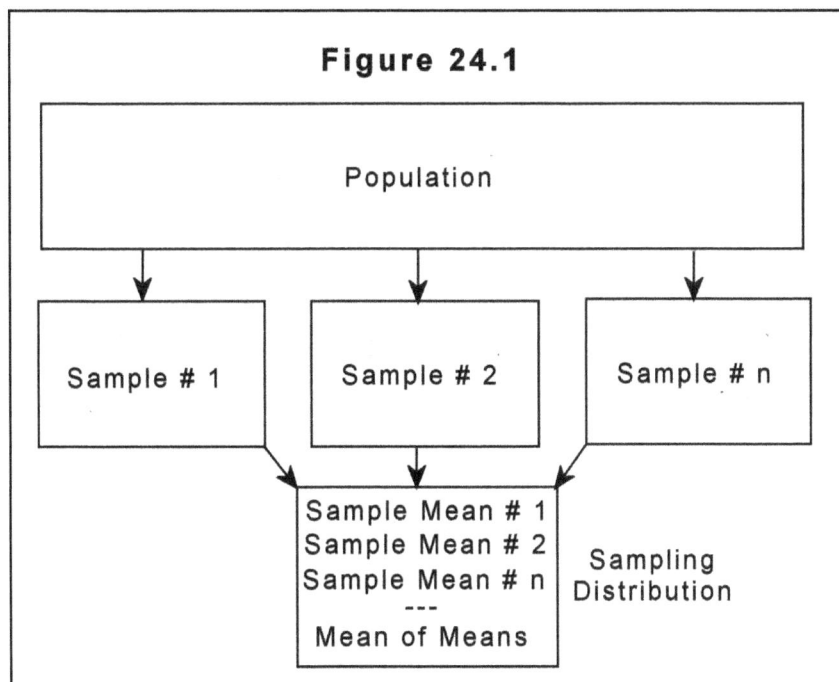

Figure 24.1

There are different sampling distributions for different sample sizes. Again for simplicity, the focus of this chapter is on all possible samples only.

Table 24.1 on Page 96 shows a population made up of the numbers 1, 2, 3, 4, and 5.

Table 24.1	
X	**Probability**
1	1/5
2	1/5
3	1/5
4	1/5
5	1/5
Total	5/5

And, Table 24.2 shows the values and the means of all possible samples of size n=2, which is 25 values, taken, with replacement, from the population shown in Table 24.1.

Table 24.2									
Value	**Mean**	**Value**	**Mean**	**Value**	**Mean**	**Value**	**Mean**	**Value**	**Mean**
1,1	1.0	2,1	1.5	3,1	2.0	4,1	2.5	5,1	3.0
1,2	1.5	2,2	2.0	3,2	2.5	4,2	3.0	5,2	3.5
1,3	2.0	2,3	2.5	3,3	3.0	4,3	3.5	5,3	4.0
1,4	2.5	2,4	3.0	3,4	3.5	4,4	4.0	5,4	4.5
1,5	3.0	2,5	3.5	3,5	4.0	4,5	4.5	5,5	5.0

Finally, looking at Table 24.2, note that a mean of 1.0 appears once, a mean of 1.5 appears twice, a mean of 2.0 appears trice, and so on and so forth. Table 24.3 on Page 97 shows the sampling distribution of the sample means of all possible samples of size n=2, which is 25 values, taken, with replacement, from the population shown in Table 24.1.

Table 24.3	
Sample Mean	Probability
1.0	1/25
1.5	2/25
2.0	3/25
2.5	4/15
3.0	5/25
3.5	4/25
4.0	3/25
4.5	2/25
5.0	1/25
Total	25/25

Shape

Figure 24.2 shows a histogram of the population shown in Table 24.1 on Page 96.

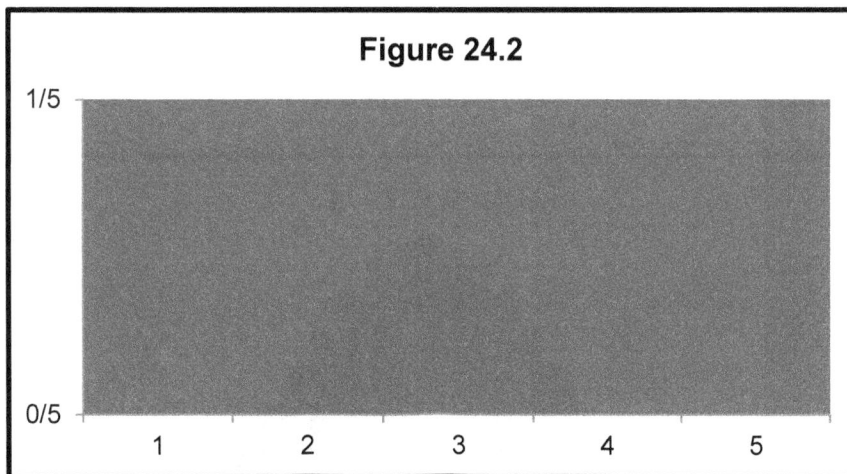

Figure 24.2

And, Figure 24.3 on Page 98 shows a histogram of the sampling distribution shown in Table 24.3.

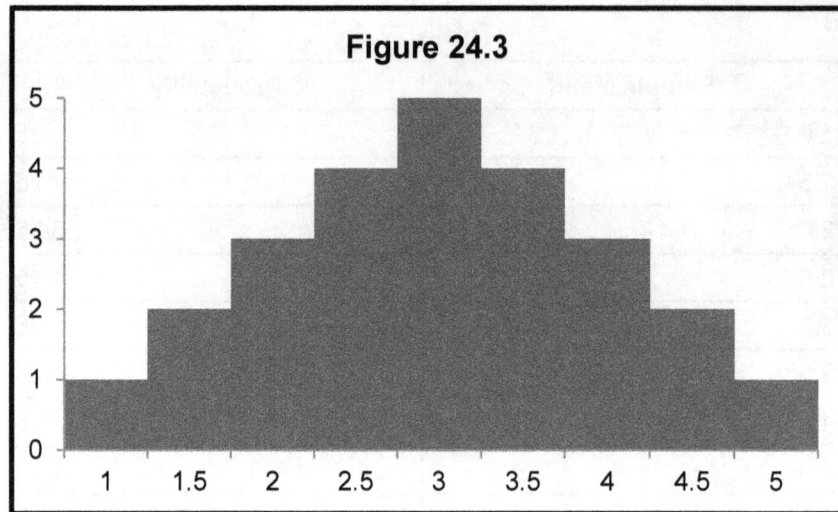

Figure 24.3

Comparing Figure 24.2 on Page 97 and Figure 24.3, note that, although the shape of the distribution of the population is uniformly distributed, the shape of the sampling distribution is "normal-like".

It can be shown that, in sampling distributions of selected random samples of size n, as the number of samples increases, the shape of the sampling distribution approaches the normal distribution (See Figure 24.4). Furthermore, it can also be shown that, as the sample size increases, not only the shape of the sampling distribution approaches the normal distribution more rapidly, but the spread of the shape of the sampling distribution decreases (See Figure 24.5 on Page 99).

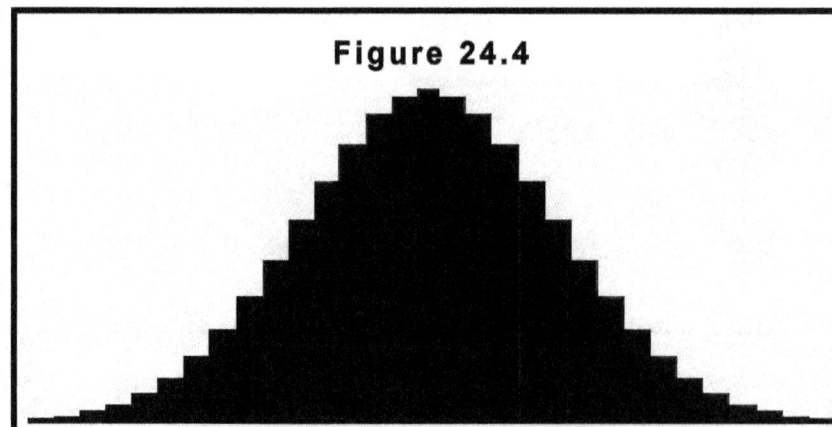

Figure 24.4

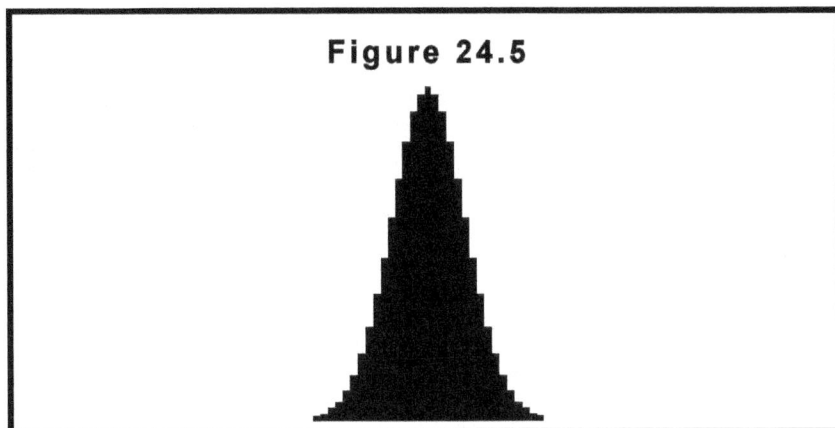

Figure 24.5

Mean

The mean of the population, μ, shown in Table 24.1 on Page 96 is 3.00:

$$(1*1/5)+(2*1/5)+...+(5*1/3)=3.00$$

And, the mean of the sampling distribution of the sample means, $\mu_{x\text{-bar}}$, shown in Table 24.3 on Page 97 —the "mean of means"— is 3.00:

$$(1*1/25)+(1.5*2/25)+...+(5*2/25)=3.00$$

Note that the mean of the sampling distribution of the sample means of all possible samples equals the population mean.

It can be shown that, in sampling distributions of selected random samples of size n, as the number of samples increases or, more importantly, as the sample size increases, the mean of the sampling distribution of the sample means approximates the population mean.

Standard Error

The standard deviation, σ, of the population shown in Table 24.1 on Page 96 is 1.4142:

$$SQRT(((1-3)^2*(1/5))+((2-3)^2*(1/5))+..+((5-3)^2*(1/5)))=1.4142$$

And, the standard deviation of the sampling distribution of the sample means, $\sigma_{x\text{-bar}}$, shown in Table 24.3 on Page 97 —called the standard error of the mean— is 1.000:

$$SQRT(((1-3)^2*(1/25))+((1.5-3)^2*(2/25))+..+((5-3)^2*(1/25)))=1.0000$$

Note that the standard deviation, σ, is a measure of the variability of the values of the population whereas the standard error of the mean, $\sigma_{x\text{-bar}}$, is a measure of the variability of the mean of the sampling distribution of the sample means.

The standard error of the mean is also equal to ...

$$\sigma_{\bar{x}} = \frac{\sigma}{\sqrt{n}}$$

... where $\sigma_{x\text{-bar}}$ is the standard error of the mean, σ is the standard deviation of the population, and n is the sample size. For example:

$$1.4142/SQRT(2)=1.0000$$

It can be shown that, in sampling distributions of selected random samples of size n, the standard error of the mean is smaller than the standard deviation of the population. Furthermore, it can also be shown that, as the sample size increases, the standard error of the mean decreases.

The standard error of the sample mean is an estimate of the standard error of the population mean and is equal to ...

$$SE_{\bar{x}} = \frac{s}{\sqrt{n}}$$

... where $SE_{x\text{-bar}}$ is the standard error of the sample mean, s is the standard deviation of the sample, and n is the sample size.

Implications

Sampling distributions are the foundation of the Central Limit Theorem.

Central Limit Theorem

Chapter 24 laid the foundation for this chapter. About the shape of distributions, recall that:

- Regardless of the shape of the distribution of the population, as the number of samples increases, the shape of the sampling distribution of the sample means approximates the normal distribution.

- And, as the sample size increases, not only the shape of the distribution approximates the normal distribution more rapidly, but the spread of the shape of the distribution decreases.

About the mean, recall that:

- The mean of the sampling distribution of the sample means , $\mu_{x\text{-bar}}$, of all possible samples equals the population mean, μ.

- As the number of samples increases the mean of the sampling distribution of the sample means, $\mu_{x\text{-bar}}$ —the "mean of means"— approximates the population mean, μ.

- And, as the sample size increases, the sampling mean, $\mu_{x\text{-bar}}$, approximates the population mean, μ, more rapidly.

And, about the standard error of the mean, recall that:

- The standard deviation of the sampling distribution of the sample means, $\sigma_{x\text{-bar}}$, which is called the standard error of the mean is a measure of the variability of the mean.

- The standard error of the mean, $\sigma_{x\text{-bar}}$, is smaller than the standard deviation of the population, σ. In fact, the standard error of the mean is equal to the standard deviation of the population divided by the square root of the sample size and is, necessarily, smaller. It follows that, as the sample size increases, the standard error of the mean decreases.

- Furthermore, the standard error of the sample mean is an estimate of the standard error of the population mean.

The Central Limit Theorem

The Central Limit Theorem (CLT) states that:

- The shape of the sampling distribution of the sample means approximates a normal distribution if either the population is normally distributed or if the sample size is large, that is, n>30.

- Furthermore, the mean of the sampling distribution, $\mu_{x\text{-bar}}$, approximates the mean of the population, μ.

- Finally, the standard error of the mean, $\sigma_{x\text{-bar}}$, is equal to the standard deviation of the population, σ, divided by the sample size, n.

Significance

Based on the CLT, one can say that:

- The shape of the distribution of a sample approximates a normal distribution if either the population is normally distributed or if the sample size is large –that is, n>30.

- Furthermore, the mean of the sample, x-bar, approximates the mean of the population, μ.

- Finally, the standard error of the sample mean, $SE_{x\text{-bar}}$, is equal to the standard deviation of the sample, s, divided by the square root of the sample size, n.

The significance of the CLT is that, based on sample statistics, one can make statements about the population parameters from which the sample was drawn.

Introduction to Estimation

Recall that parameters are values that describe the characteristics of populations whereas statistics are values that describe the characteristics of samples. Also recall that, oftentimes, the parameters of populations are unknown and the statistics of samples are used to estimate the parameters of populations. For example, one might estimate the average weight of the U.S. population by weighing a portion of the population. Finally, recall that estimation includes point estimation and interval estimation.

Point Estimates

A point estimate is a single value that describes a characteristic of a population based on a characteristic of a random sample drawn from the population. For example, the sample mean, a statistic, is an estimate of the population mean, a parameter.

Interval Estimates

An interval estimate —called a confidence interval (CI)— is a range that measures of the reliability of a sample statistic. A confidence interval is made up of a point estimate, for example, a sample mean, and a margin of error. The confidence interval is …

$$\text{Confidence Interval} = \text{Point Estimate} \pm \text{Margin of Error}$$

Figure 26.1 shows the confidence interval (CI).

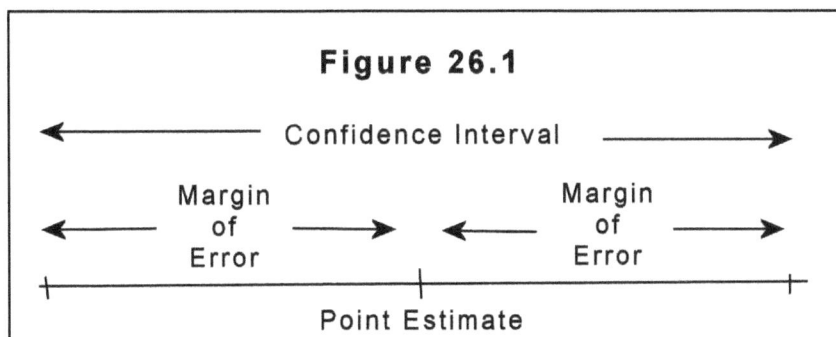

Figure 26.1

Margin of Error. The margin of error is made up of a measure of the variability of the sample statistic, for example, a standard error of the sample mean, and a value, for example, a z score —called a critical value— that corresponds to a selected level of confidence (CL) —that is, a probability. The margin of error is …

Margin of Error=Variability * Critical Value

Critical Values and Confidence Level. Tables, for example, a z table, are used for finding the critical values that correspond to a level of confidence (CL), where CL=100%-α. For example, Table 26.1 and Figure 26.2 show the critical values for the 95% confidence level, that is, CL=100%-5%. Note that α/2=2.5%.

Table 26.1		
z Score		Cumulative Left-Tail Probability
-1.9	.60	0.02500
-1.6	.45	0.05000
-1.2	.82	0.10000
+1.2	.82	0.90000
+1.6	.45	0.95000
+1.9	.60	0.97500

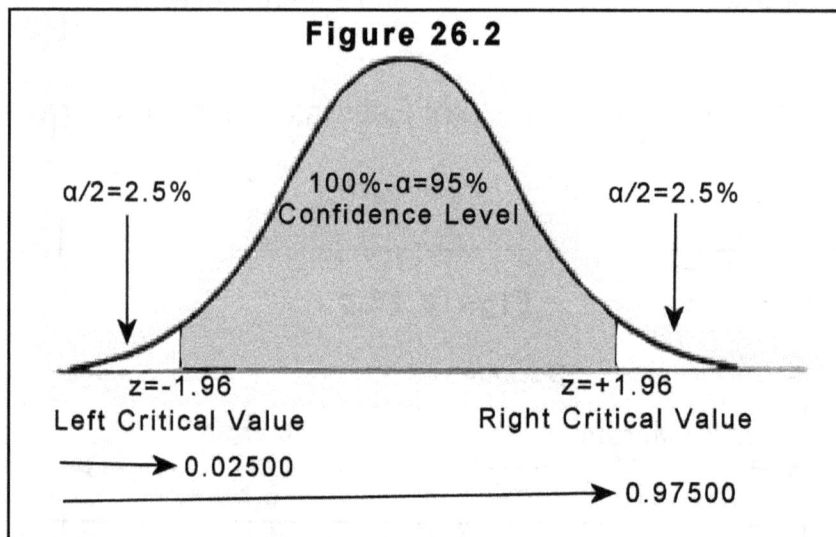

Figure 26.2

α/2=2.5%   100%-α=95% Confidence Level   α/2=2.5%

z=-1.96
Left Critical Value

z=+1.96
Right Critical Value

⟶ 0.02500

⟶ 0.97500

Caveat! A confidence interval (CI) shows the range of the sample statistic, not the range of the population parameter. Why? Because population parameters do not vary. However, a confidence interval (CI) does show the range within the true population parameter is found.

Estimating Means

Recall that the sample mean, x-bar, a statistic, is an estimate of the population mean, μ, a parameter. Also recall that estimation includes point estimation and interval estimation.

Sample Data

The population made up of the heights of all adult males in the United States is normally distributed with a mean of 70 inches and a standard deviation of 3 inches, which, for the purposes of this chapter, are unknown. Say one draws a random sample of size 25 (See Table 27.1).

Table 27.1				
71.0200	71.6461	68.5963	67.9120	69.0970
71.4197	67.6320	72.2335	72.5737	71.0843
70.3745	71.0686	72.0944	67.7740	72.3278
71.5249	70.7777	69.2637	65.5464	73.3561
71.9407	71.9695	68.6985	64.7875	66.9267

Point Estimate

Recall that a point estimate is a single value that describes a characteristic of a population based on a characteristic of a random sample drawn from the population.

Sample Mean. The sample mean, x-bar, of the distributions shown in Table 27.1 is 70.07 inches:

$$(71.0200+71.4197+...+66.9267)/25=70.07$$

Interval Estimate

Recall that an interval estimate —called a confidence interval (CI)— is a range that measures the reliability of a sample statistic. Also recall that a confidence interval is made up of:

- A point estimate, for example, a sample mean.
- A measure of the variability of the sample statistic, for example, a standard error of the mean.
- A value —called a critical value— for example, a z score, that corresponds to a selected level of confidence (CL) —that is, a probability.

In this example, because the population standard deviation, σ, is unknown, one uses the standard error of the sample mean, $SE_{x\text{-}bar}$, rather than the standard error of the mean, $σ_{x\text{-}bar}$. And, also, one uses a t score rather than a z score.

Standard Error. The standard error of the sample mean, $Se_{x\text{-}bar}$, is a measure of the variability of the sample mean and is equal to …

$$SE_{\bar{x}} = \frac{s}{\sqrt{n}}$$

… where $SE_{x\text{-}bar}$ is the standard error of the sample mean, s is the standard deviation of the sample, and n is the sample size.

The standard error of the sample mean of the distribution shown in Table 27.1 on Page 105 is 0.4615:

SQRT((($71.02-70.07)^2+(71.42-70.07)^2+...+(66.93-70.07)^2$)/24) /SQRT(25)=0.4615

Critical Values and Confidence Level. The t table is used for finding the critical values (CVs) of t that correspond to a selected level of confidence (CL), where CL=100%-α, which, for this example, 90% is selected. Table 27.1 and Figure 27.1 on Page 107 show, for DF=24, that the t scores that correspond to the 90% confidence level, which is CL=100%-10%, are -1.711 for the left side and +1.711 for the right side. The confidence level shows the probability, which is 90%, that the t scores are between -1.711 and +1.711. Note that α/2=5.0%.

Table 27.2			
Degrees of Freedom	**Probability (One Tail)**		
	0.1000	0.0500	0.0250
	1.318	1.711	2.064
24	**Probability (Two Tail)**		
	0.1000 (0.0500+0.0500)	0.0500 (0.0250+0.0250)	0.0250 (0.0125+0.0125)
	1.711	2.064	2.391

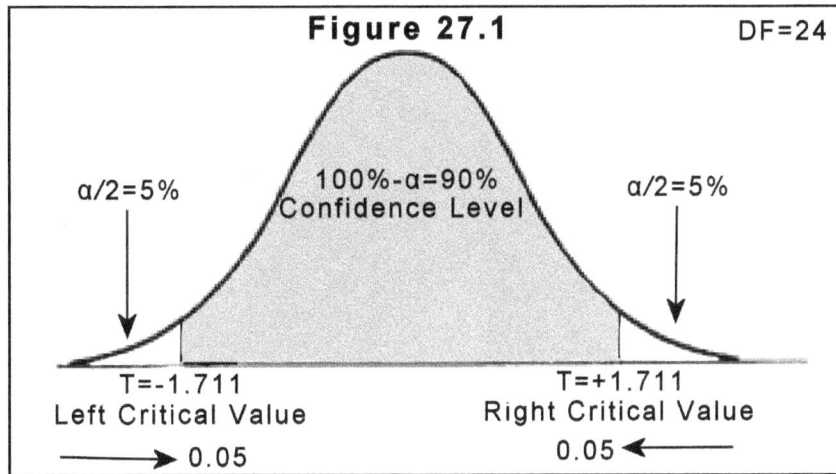

Figure 27.1 DF=24

α/2=5%

100%-α=90%
Confidence Level

α/2=5%

T=-1.711
Left Critical Value
⟶ 0.05

T=+1.711
Right Critical Value
0.05 ⟵

Confidence Interval. The 90% confidence interval of the sample mean of the distribution shown in Table 27.1 on Page 105 is:

$$70.07\pm(0.4615*1.711)=69.28 \text{ inches to } 70.86 \text{ inches}$$

That is, in 90% of the times, the sample mean will be between a lower boundary (LB) of 69.28 inches and an upper boundary (UB) of 70.86 inches (See Figure 27.2).

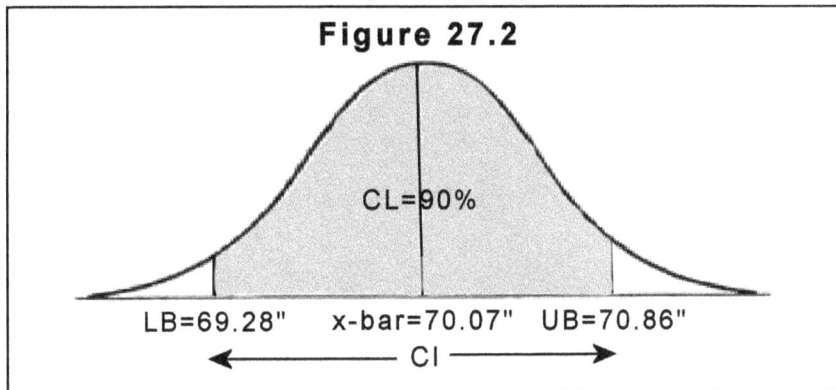

Figure 27.2

CL=90%

LB=69.28" x-bar=70.07" UB=70.86"
⟵ CI ⟶

Caveat! The confidence interval does not state that, in 90% of the times, the population mean will be between 69.28" and 70.86". Why? Because the population mean does not vary. However, one may say that one is "confident" that, in 90% of the times, the true value of the population mean is within the confidence interval.

Blank Page

Estimating Proportions

Recall that proportions describe the relationship between the frequencies of values and the whole. For example, about 2 out 5, or 2/5, or 0.40, or 40% of all adult Americans are college grads. Also recall that estimation includes point estimation and interval estimation. The ways for estimating proportions parallel the ways for estimating means.

Sample Data

In a randomized survey, 385 out of 1,000 adult Americans said that they graduated college.

Point Estimate

Recall that a point estimate is a single value that describes a characteristic of a population based on a characteristic of a random sample drawn from the population. The sample proportion, p-hat, is the best estimate of the population proportion, p. The formula for the sample proportion is …

$$\text{p-hat} = \frac{x}{n}$$

… where p-hat is the sample proportion, x is the count of a characteristic, and n is the sample size.

Sample Proportion. The sample proportion for the above-referenced survey is 0.385:

$$385/1{,}000 = 0.385$$

Interval Estimate

Recall that an interval estimate —called a confidence interval (CI)— is a range that measures the reliability of a sample statistic. Also recall that a confidence interval is made up of:

- A point estimate, for example, a sample proportion.
- A measure of the variability of the sample statistic, for example, a standard error of the mean.
- A value —called a critical value— for example, a z score, that corresponds to a

selected level of confidence (CL) —that is, a probability.

This section describes how to estimate the 90% confidence interval (CI) of a proportion using the sample data from the above-referenced survey.

Standard Error. The standard error of the sample proportion, $SE_{p\text{-hat}}$, is an estimate of the standard error of the population proportion, SE_p. The formula for the standard error of the sample proportion is …

$$SE_{p\text{-hat}} = \sqrt{\frac{p\text{-hat} * (1 - p\text{-hat})}{n}}$$

… where $SE_{p\text{-hat}}$ is the standard error of the sample proportion, p-hat is the sample proportion, and n is the sample size. Additionally, np(1-p)>10 and n≤0.05N.

The standard error of the sample proportion for the above-referenced survey is 0.0154:

$$\text{SQRT}((0.385*(1-0.385))/1{,}000)=0.0154$$

Critical Values and Confidence Level. The z table is used for finding the critical values (CVs) of z that correspond to a selected level of confidence (CL), where CL=100%-α, which, for this example, 90% is selected. Table 28.1 and Figure 28.1 on Page 111 show that the z scores that correspond to the 90% confidence level, which is CL=100%-10%, are -1.645 for the left side and +1.645 for the right side. The confidence level shows the probability, which is 90%, that the z scores are between -1.645 and +1.645. Note that α/2=5.0%.

Table 28.1		
z Score		**Cumulative Left-Tail Probability**
-1.9	.60	0.02500
-1.6	.45	0.05000
-1.2	.82	0.10000
+1.2	.82	0.90000
+1.6	.45	0.95000
+1.9	.60	0.97500

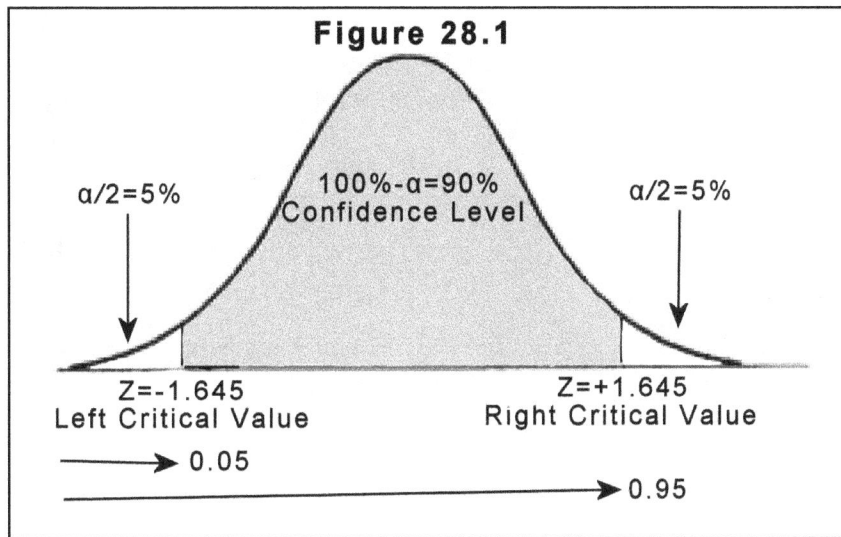

Figure 28.1

Confidence Interval. The 90% confidence interval of the sample proportion for the above-referenced survey is:

$$0.385 \pm (0.0154*1.645) = 0.3597 \text{ to } 0.4103$$

That is, in 90% of the times, the sample proportion will be between a lower boundary (LB) of 35.97% and an upper boundary (UB) of 41.03% (See Figure 28.2).

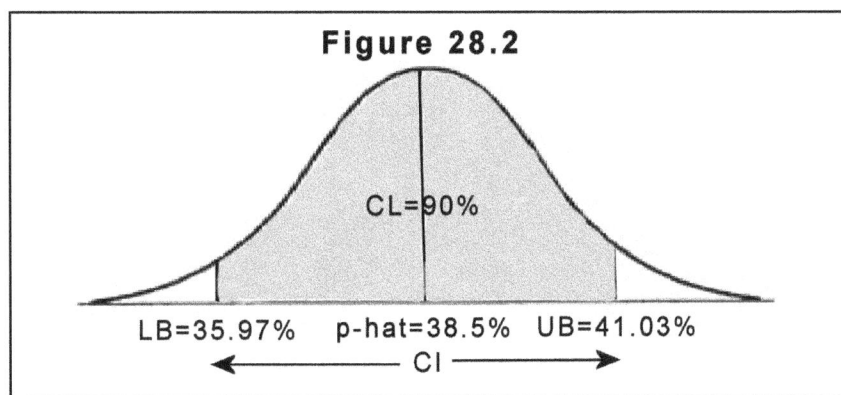

Figure 28.2

Caveat! The confidence interval does not state that, in 90% of the times, the population proportion will be between 35.97% and 41.03%. Why? Because the population proportion does not vary. However, one may say that one is "confident" that, in 90% of the times, the true value of the population proportion is within the confidence interval.

Small Populations

For small populations, the formula for the standard error of the sample proportion is ...

$$SE_{p\text{-hat}} = \sqrt{\frac{p\text{-hat}*(1 - p\text{-hat})}{n} * \frac{N-n}{N-1}}$$

... where $SE_{p\text{-hat}}$ is the standard error of the sample proportion, p-hat is the sample proportion, n is the sample size, and N is the population size.

Estimating Variances

Recall that variances are measures of dispersion. Also recall that estimation includes point estimation and interval estimation. The ways for estimating variances somewhat parallel the ways for estimating means and proportions. However, there are two differences:

- Means and proportions follow a normal distribution whereas variances follow a chi-square distribution. Therefore, the left-side and the right-side critical values are not equal (See Figure 29.1).

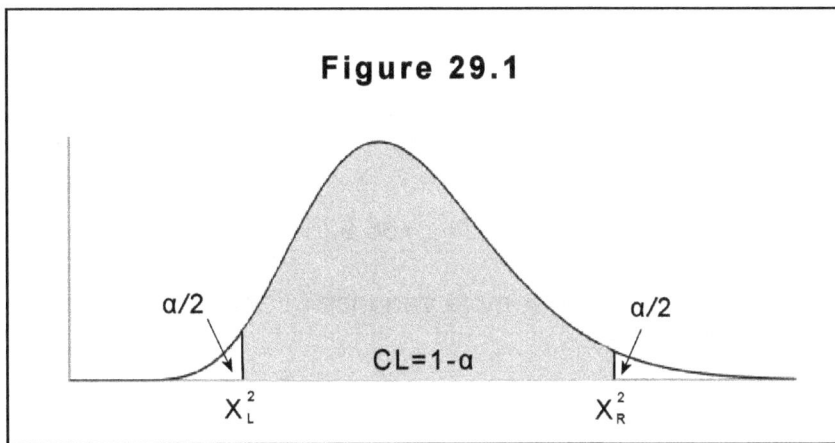

Figure 29.1

- There is a way for directly calculating the boundaries of the confidence interval (CI).

Sample Data

The population made up of the heights of all adult males in the United States is normally distributed with a mean of 70 inches and a standard deviation of 3 inches, which, for the purposes of this chapter, are unknown. Say one draws a random sample of size 25 (See Table 29.1).

Table 29.1				
71.0200	71.6461	68.5963	67.9120	69.0970
71.4197	67.6320	72.2335	72.5737	71.0843
70.3745	71.0686	72.0944	67.7740	72.3278
71.5249	70.7777	69.2637	65.5464	73.3561
71.9407	71.9695	68.6985	64.7875	66.9267

Point Estimate

Recall that a point estimate is a single value that describes a characteristic of a population based on a characteristic of a random sample drawn from the population. The sample variance, s^2, is the best estimate of the population variance, σ^2. The formula for the sample variance is ...

$$s^2 = \frac{\sum (x-\bar{x})^2}{n-1}$$

... where s^2 is the sample variance, x are the sample values, x-bar is the sample mean, and n is the sample size.

Sample Mean. The sample mean for the distribution shown in Table 29.1 on Page 113 is 70.07 inches:

$$(71.0200+71.4197+...+66.9267)/25=70.07$$

Sample Variance. And, the sample variance for the distribution shown in Table 29.1 on Page 113 is 5.32 inches:

$$((71.02-70.07)^2+(71.42-70.07)^2+...+(66.93-70.07)^2)/24=5.32$$

Interval Estimate

Recall that an interval estimate —called a confidence interval (CI)— is a range that measures the reliability of a value. This section describes how to estimate the 95% confidence interval of a population variance using the above-referenced sample data.

Critical Values and Confidence Level. The chi-square table is used for finding the critical values (CVs) of chi-square that correspond to a selected level of confidence (CL), where CL=100%-α, which, for this example, 95% is selected.

Table 29.2 on Page 115 and Figure 29.2 on Page 115 show that, for DF=24, the chi-square statistics that correspond to the 95% confidence level, which is CL=100%-5%, are 12.401 for the left side and 39.364 for the right side. The confidence level shows the probability, which is 95%, that the chi-square statistics are between 12.401 and 39.364. Note that α/2=2.5%.

Table 29.2		
Degrees of Freedom	Probability (Right-Tail)	
	0.975	0.025
24	12.401	39.364

Figure 29.2

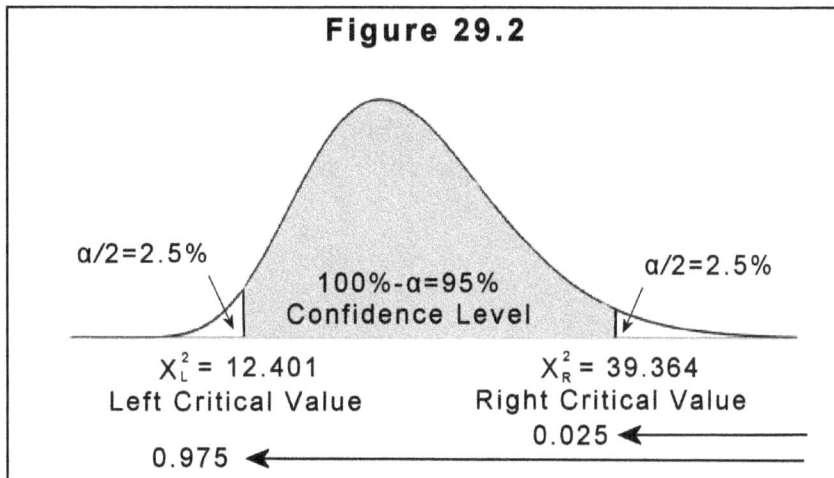

α/2=2.5%

100%-α=95% Confidence Level

α/2=2.5%

X_L^2 = 12.401
Left Critical Value

X_R^2 = 39.364
Right Critical Value

0.025 ←

0.975 ←

Confidence Interval. The formula for calculating the boundaries of the population variance is ...

$$LB/UP = \frac{(n-1)*s^2}{X^2}$$

... where LB/UP is the lower or upper boundary, n is the sample size, s^2 is the sample variance, and X^2 is the chi-square statistic.

The 95% confidence interval of the population variance of the distribution shown in Table 29.1 on Page 113 is:

The lower boundary (LB) of the confidence interval is 3.24:

$$(25-1)*5.32/39.364=3.24$$

And, the upper boundary (UB) of the confidence interval is 10.30:

$$(25-1)*5.32/12.401=10.30$$

That is, in 95% of the times, the population variance will be between 3.24 and 10.30. Note that the right critical value is used for calculating the lower boundary and the left critical value is used for calculating the upper boundary.

Caveat! The confidence interval does not state that, in 95% of the times, the population variance will be between 3.24 and 10.30. Why? Because the population variance does not vary. However, one may say that one is "confident" that, in 95% of the times, the true value of the population variance is within the confidence interval.

Confidence Interval of the Sample Standard Deviation

The confidence interval of the sample standard deviation is simply the square root of the confidence interval of the sample variance.

Introduction to Hypothesis Testing

Recall that a hypothesis is a statement about a characteristic of a population — that is, the variable to be tested. For example, one may say that the life expectancy of Hispanic American males is about 80 years. Also recall that, simply stated, hypothesis tests, also called significance tests, are ways for failing to reject —that is, accepting— or rejecting such statements.

Hypothesis tests might compare a population parameter to a hypothesized value or might compare two or more population parameters. For simplicity, the focus of this chapter is on comparing a population parameter to a hypothesized value.

Null and Alternative Hypotheses

The null hypothesis, labeled H_0, says that a population parameter, for example, a mean, is equal to a stated value. For example, the average height of adult males in the United States is equal to 5' 10", that is, H_0: $\mu=70$". In other words, the null hypothesis says that there is no difference between the average height and 5' 10". Thus, the null hypothesis is the "no difference" hypothesis.

The alternative hypothesis, labeled H_1, says that the population parameter, for example, the mean, is not equal to, is more than, or is less than the stated value. For example:

- The average height is not equal to 5' 10", that is, H_1: $\mu \neq 70$"
- Or the average height is more than 5'10", that is, H_1: $\mu>70$"
- Or the average height is less than 5'10", that is, H_1: $\mu<70$"

That is, there is a difference between the average height and 5' 10". Table 30.1 shows examples of the null and the alternative hypotheses.

Table 30.1		
	H_0	H_1
Bi-directional	$\mu=70$"	$\mu \neq 70$"
Directional (Right- tailed)	$\mu=70$"	$\mu>70$"
Directional (Left-tailed)	$\mu=70$"	$\mu<70$"

In bi-directional tests, the alternative hypothesis simply says that there is a difference between the population parameter and the stated value. In directional tests, the alternative hypothesis says that population parameter is either more than or less than the stated value.

Hypothesis Testing

       Hypothesis tests are ways for either failing to reject —that is, accepting— or rejecting the null hypothesis. Generally speaking, one may say that hypothesis tests are ways of assigning probabilities to the location of the sample statistic vis-à-vis the location of the population parameter.

       For example, for a right-tailed test, the grayed area of Figure 30.1 shows the probability that the sample statistic is to the right of the population parameter and the blacked area shows the probability that the sample statistic is farther to the right of the population parameter. The grayed is called the acceptance region and the blacked area is called the rejection region —called the critical region.

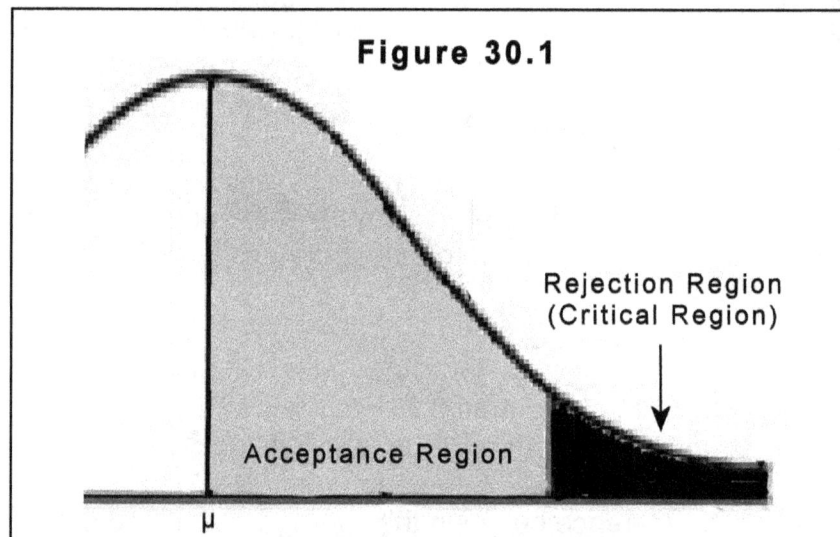

Figure 30.1

Rejection Region
(Critical Region)

Acceptance Region

μ

Example

       Say that one states that the average SAT math scores of male students is 500 and that the scores are normally distributed. Then, say that one draws three random samples from the population. Most likely, the three samples yield different means (See Figure 30.2 on Page 119).

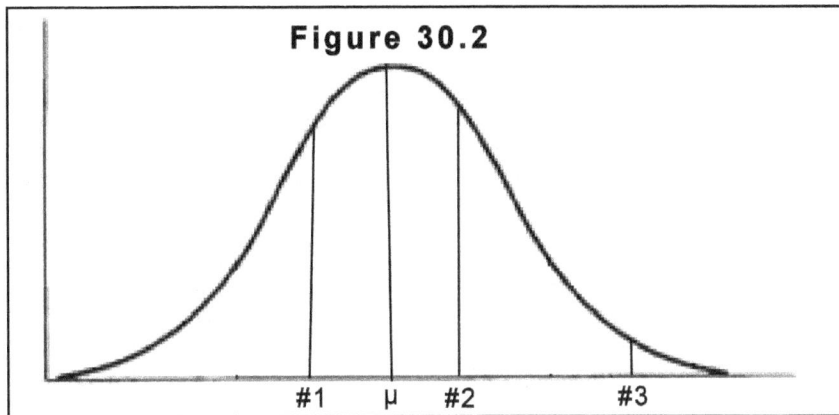

Figure 30.2

Looking at the means of samples # 1 and # 2 alone, intuitively, there is no reason to believe that the population mean is not 500. However, looking at sample # 3 alone, intuitively, there is reason to believe that the population mean might not be 500. In fact, if the population mean is 500, the probability of a sample mean being so far to the right, while not impossible, is very small (See Figure 30.3). Therefore, one might infer, rightly or wrongly, that the population mean is not 500.

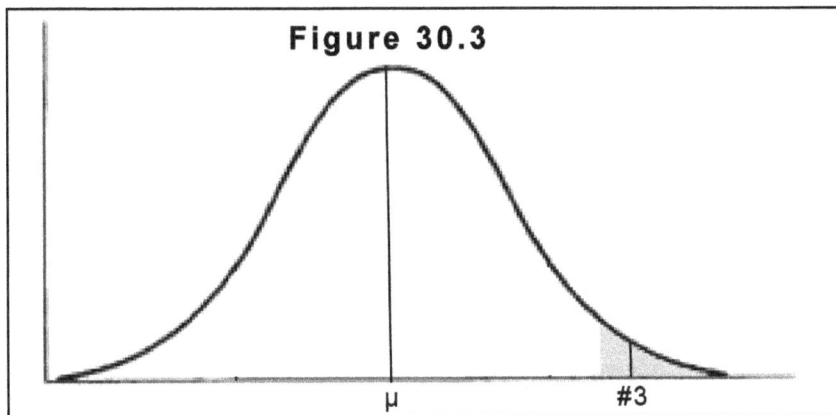

Figure 30.3

Elements of Hypothesis Testing

Figure 30.4 on Page 120 shows the elements of hypothesis testing:

- **Confidence Level.** The confidence level (CL), shown in gray, is the probability that the sample statistic is located within the acceptance region.

- **Alpha.** Alpha (α) —called the significance level— shown in black, is the probability that the sample statistic is located within the rejection region —called. the critical region.

Note that the confidence level is equal to to 1 – alpha. Likewise, alpha is equal to 1 – the confidence level.

- **Critical Values.** The critical values (CVs) are the values, for example, z scores, that divide the acceptance region from the rejection region —called the critical region. Critical values are contingent on the confidence level.

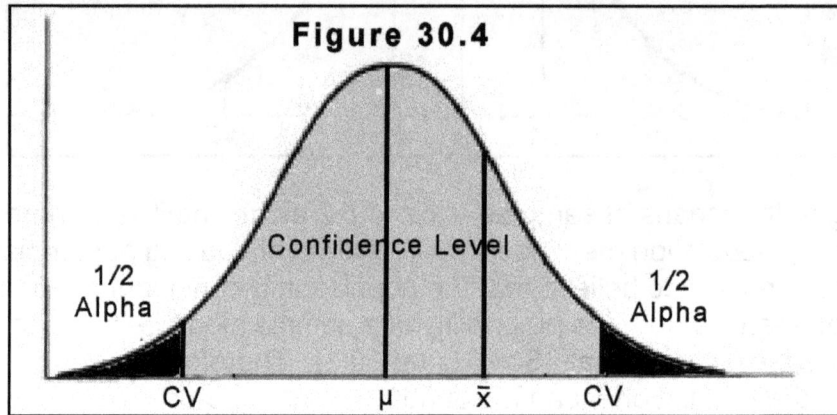

Figure 30.4

- **P-Value.** The p-value is a probability that is compared to the significance level.

Methodology

First, review the methodology described in Chapter 23. This describes the steps to be followed for a one-sample t test:

- State the null hypothesis and the alternative hypothesis.
- Calculate the sample mean.
- Calculate the sample standard deviation.
- Calculate the relevant t score.
- Select a confidence level.
- Find the critical values that correspond to the selected confidence level.
- Fail to reject —that is, accept— or reject the null hypothesis.
- Alternatively, calculate the p-value.

Errors

If the null hypothesis is true but, by chance, the sample statistic is located within the rejection region, that is, the critical region, one would incorrectly reject the null hypothesis and make a Type I Error. Conversely, if the null hypothesis is false but, by chance, the sample statistic is located within the acceptance region, one would incorrectly fail to reject —that is, accept— the null hypothesis and make a Type II Error. Table 30.2 shows the error types.

Table 30.2		
	H_0 **True**	H_0 **False**
Reject H_0	Type I Error	Right Inference
Fail to Reject H_0	Right Inference	Type II Error

Parametric Tests versus Non-parametric Tests

Again review the methodology described in Chapter 23. Say that one is testing either a qualitative variable or a quantitative variable that is not normally distributed and that the sample size is small, that is, $n < 30$. Under these conditions, one cannot use typical hypothesis tests, for example, one-sample t tests; tests that are called parametric tests. Under these conditions, one uses other types of tests; tests that are called non-parametric tests.

Blank Page

Testing a Hypothesized Mean

This chapter describes a t test for comparing a sample mean, a statistic, which is a proxy for a population mean, a parameter, to a hypothesized population mean. This test is known as a one-sample t test. First, review the methodology described in Chapter 23. Then, from Chapter 30 recall that the steps to be followed for a one-sample t test are:

- State the null hypothesis and the alternative hypothesis.
- Calculate the sample mean.
- Calculate the sample standard deviation.
- Calculate the relevant t score.
- Select a confidence level.
- Find the critical values that correspond to the selected confidence level.
- Fail to reject —that is, accept— or reject the null hypothesis.
- Alternatively, calculate the p-value.

Sample Data

SAT scores are approximately normally distributed. Table 31.1 shows a random sample of the SAT scores of 25 male students.

Table 31.1				
SAT Scores				
504	519	462	444	530
347	538	561	709	463
640	427	450	538	469
410	521	520	494	392
369	431	487	361	506

Null Hypothesis and Alternative Hypothesis

Say that one states that the average SAT score is equal to 500. That is, H_0: $\mu=500$ and H_1: $\mu\neq500$. Is it true?

Figure 31.1 on Page 124 shows that, if the the test shows that the calculated t score is within the acceptance region, shown in gray, then, one fails to reject —that is, accepts— the null hypothesis and, therefore, is likely that there is no difference between the average SAT scores of the students and 500. Conversely, if the test shows that the

the calculated t score is within the rejection regions, shown in black, then, one rejects the null hypothesis and, therefore, is likely that there is a difference between the average SAT scores of the students and 500.

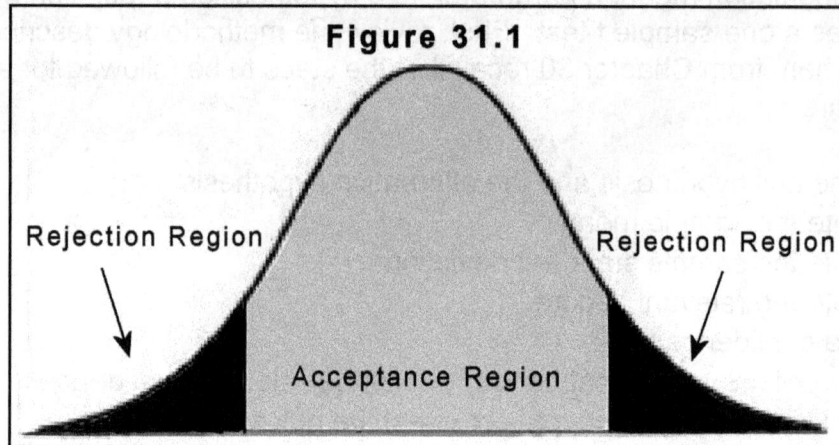

Figure 31.1

Sample Mean

The sample mean of the distribution shown in Table 31.1 on Page 123 is 483.64:

$$(504+347+...+506)/25=483.64$$

Sample Standard Deviation

The sample standard deviation of the distribution shown in Table 31.1 on Page 123 is 82.86:

$$SQRT(((504-483.64)^2+(347-483,64)^2+...+(506-483.64)^2)/24)=82.86$$

T Score

Recall that the formula for calculating the t score of a sample mean is ...

$$t=\frac{\bar{x}-\mu}{\frac{s}{\sqrt{n}}}$$

... where t is the t score, x-bar is the sample mean, μ is the population mean, which is the hypothesized mean, s is the standard deviation of the sample, and n is the sample size.

The relevant t score is -0.987:

$$(483.64-500)/(82.86/SQRT(25))=-0.987$$

Degrees of Freedom

For this example, the degrees of freedom are 25-1=24.

Confidence Level and Critical Values

The t table is used for finding the critical values (CVs) of t that correspond to a selected level of confidence, where CL=100%-α, which, for this example, 90% is selected. Table 31.2 and Figure 31.2 on Page 126 show, for DF=24, that the t scores that correspond to the 90% confidence level, which is CL=100%-10%, are -1.711 for the left side and +1.711 for the right side. The confidence level shows the probability, which is, 90%, that the t scores are between -1.711 and +1.711. Note that α/2=5.0%.

Table 31.2			
Degrees of Freedom	**Probability (One Tail)**		
	0.1000	0.0500	0.0250
	1.318	1.711	2.064
	Probability (Two Tail)		
24	0.1000 (0.0500+0.0500)	0.0500 (0.0250+0.0250)	0.0250 (0.0125+0.0125)
	1.711	2.064	2.391

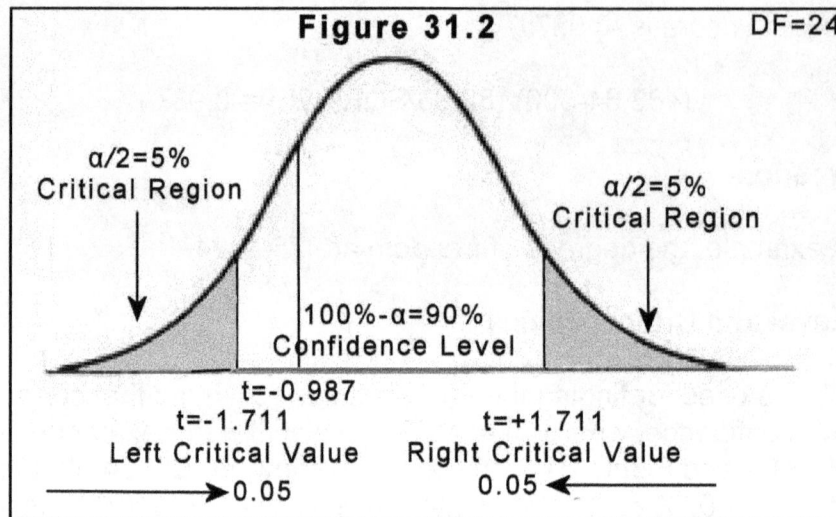

Figure 31.2   DF=24

α/2=5%
Critical Region

α/2=5%
Critical Region

100%-α=90%
Confidence Level

t=-0.987
t=-1.711
Left Critical Value
0.05

t=+1.711
Right Critical Value
0.05

Inference

Figure 31.2 shows that, for DF=24, the t score of -0.987 is between the critical values of -1.711 and +1.711 and within the 90% confidence level. Therefore, one does not reject the null hypothesis. That is, one can say that, most likely, the average SAT score is within the 90% confidence level.

Z Test

Again review the methodology described in Chapter 23. If the standard deviation of the population is known, then one uses a z score rather than a t score. And, according to some, but not to all, if the sample size is greater than 30, then one also uses a z score rather than a t score.

P Value

As an alternative to the traditional test, one may select an alpha level, say α=10%, and calculate the p=value. If p>α one fails to reject —that is, accepts— H_0. For example, using a web-based calculator, for DF=24, the two-tail probability of a t score of -0.987 is 0.333 (See Figure 31.3). [1] And, p=0.333 is more than 10%. Thus, one does not reject H_0.

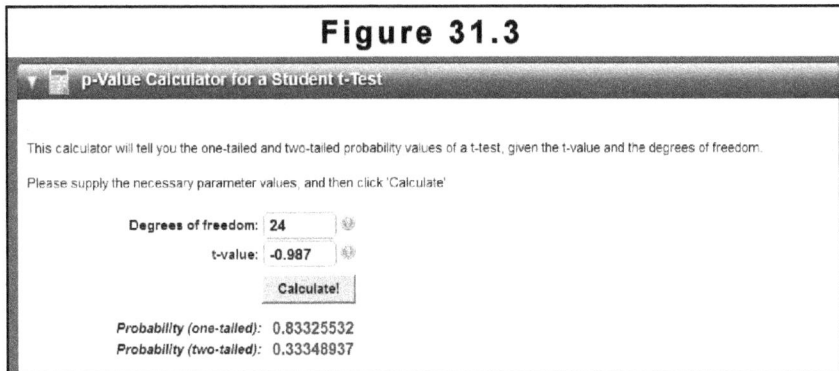

Figure 31.3

p-Value Calculator for a Student t-Test

This calculator will tell you the one-tailed and two-tailed probability values of a t-test, given the t-value and the degrees of freedom.

Please supply the necessary parameter values, and then click 'Calculate'

Degrees of freedom: 24

t-value: -0.987

Calculate!

Probability (one-tailed): 0.83325532
Probability (two-tailed): 0.33348937

Endnote

[1] Soper, D. (2015). *p-value calculator*. URL:http://www.DanielSoper.com

Blank Page

Testing Two Independent Means

Chapter 31 described a t test for comparing a sample mean, a statistic, which is a proxy for a population mean, a parameter, to a hypothesized population mean. This test is known as a one-sample t test. This chapter describes a t test for comparing the means of two independent samples, which are proxies for two population means. This test is known as a two-sample t test. To start, review the methodologies described in Chapters 23 and 30.

Sample Data

SAT scores are approximately normally distributed. Table 32.1 shows a random sample of the SAT scores of 25 male students and the SAT scores of 25 female students.

Table 32.1				
SAT Scores - Males				
504	519	462	444	530
347	538	561	709	463
640	427	450	538	469
410	521	520	494	392
369	431	487	361	506
SAT Scores - Females				
357	573	422	544	619
619	478	563	416	653
438	414	721	620	560
665	394	504	434	478
474	348	586	379	581

Null Hypothesis and Alternative Hypothesis

Say that one states that the average SAT scores of male students is equal to the average SAT scores of female students. That is, H_0: $\mu_m = \mu_f$ and H_1: $\mu_m \neq \mu_f$. Is it true?

Figure 32.1 on Page 130 shows that, if the the test shows that the calculated t score is within the acceptance region, shown in gray, then, one fails to reject —that is, accepts—the null hypothesis and, therefore, is likely that there is no difference between the average SAT scores of the male students and the average SAT scores of the female students. Conversely, if the test shows that the the calculated t score is within the rejection regions, shown in black, then, one rejects the null hypothesis and, therefore, is likely that there is a difference between the average SAT scores of the male students and the average SAT scores of the female students.

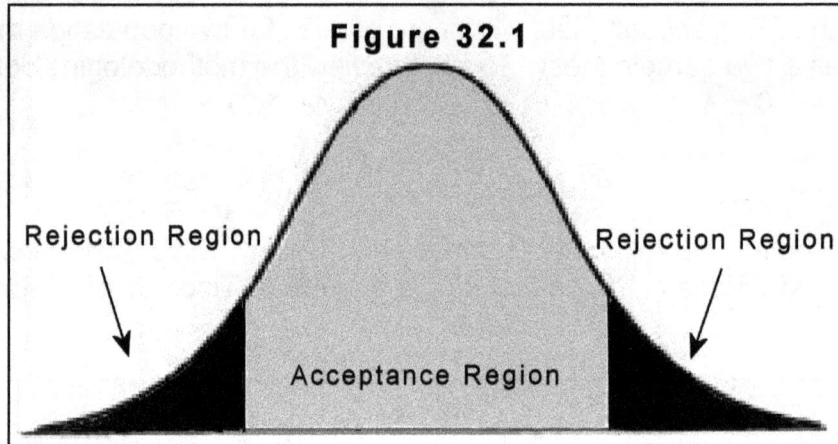

Figure 32.1

Sample Means

The sample mean of the two distributions shown in Table 32.1 on Page 129 are 483.64 and 517.60:

$$(504+347+...+506)/25=483.64$$

$$(357+619+...+581)/25=517.60$$

Sample Standard Deviations

The sample standard deviation of the two distributions shown in Table 32.1 on Page 129 are 82.86 and 104.06:

$$SQRT(((504-483.64)^2+(347-483,64)^2+...+(506-483.64)^2)/24)=82.86$$

$$SQRT(((357-517.60)^2+(619-517.60)^2+...+(581-517.60)^2)/24)=104.06$$

T Score

The formula for calculating the t score is …

$$t=\frac{(\overline{x}_1-\overline{x}_2)-(\mu_1-\mu_2)}{\sqrt{\dfrac{s_1^2}{n_1}+\dfrac{s_2^2}{n_2}}}$$

… where t is the t score, x-bar are the sample means, μ are the population means, s are the sample standard deviations, and n are the sample sizes.

For this example, $(\mu_1-\mu_2)=0$ and the relevant t score is -1.277:

(483.64-517.60-0)/SQRT((82.86^2/25)+(104.06^2/25))=-1.277

Degrees of Freedom

The formula for calculating the degrees of freedom is …

$$DF=\frac{(\dfrac{s_1^2}{n_1}+\dfrac{s_2^2}{n_2})^2}{\dfrac{(\dfrac{s_1^2}{n_1})^2}{n_1-1}+\dfrac{(\dfrac{s_2^2}{n_2})^2}{n_2-1}}$$

… where DF is the degrees of freedom, s are the standard deviations, and n are the sample sizes.

For this example, DF=46 (rounded):

((82.86^2/25+104.06^2/25)^2)/(((82.86^2/25)^2/(25 − 1))+((104.06^2/25)^2/(25-1)))=45.71

Confidence Level and Critical Values

The t table is used for finding the critical values (CVs) of t that correspond to a selected level of confidence, where CL=100%-α, which, for this example, 90% is selected. Table 32.2 on Page 132 and Figure 32.2 on Page 132 show, for DF=46, that the t scores that correspond to the 90% confidence level, which is CL=100%-10%, are -1.679 for the left side and +1.679 for the right side. The confidence level shows the probability, which is 90%, that the t scores are between -1.679 and +1.679. Note that α/2=5.0%.

Table 32.2			
Degrees of Freedom	**Probability (One Tail)**		
	0.1000	0.0500	0.0250
	1.300	1.679	2.013
46	**Probability (Two Tail)**		
	0.1000 (0.0500+0.0500)	0.0500 (0.0250+0.0250)	0.0250 (0.0125+0.0125)
	1.679	2.013	2.317

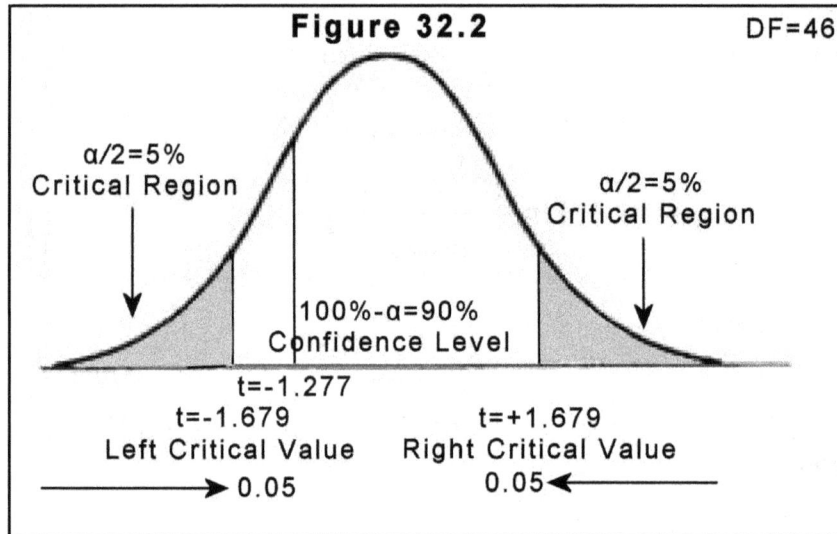

Figure 32.2   DF=46

α/2=5% Critical Region

α/2=5% Critical Region

100%-α=90% Confidence Level

t=-1.277
t=-1.679
Left Critical Value
⟶ 0.05

t=+1.679
Right Critical Value
0.05 ⟵

Inference

Figure 32.2 shows that, for DF=46, the t score of -1.277 is between the critical values of -1.679 and +1.679 and within the 90% confidence level. Therefore, one does not reject the null hypothesis. That is, one can say that, most likely, there is no difference between the average SAT scores of male students and the average SAT scores of female students.

Z Test

Again review the methodology described in Chapter 23. If the standard deviation of the population is known, then one uses a z score rather than a t score. And, according to some, but not to all, if the sample size is greater than 30, then one also uses a z score rather than a t score.

P Value

As an alternative to the traditional test, one may select an alpha level, say $\alpha=10\%$, and calculate the p=value. If $p>\alpha$ one fails to reject —that is, accepts— H_0. For example, using a web-based calculator, for DF=46, the two-tail probability of a t score of -1.277 is 0.208 (See Figure 32.3). [1] And, p=0.208 is more than 10%. Thus, one does not reject H_0.

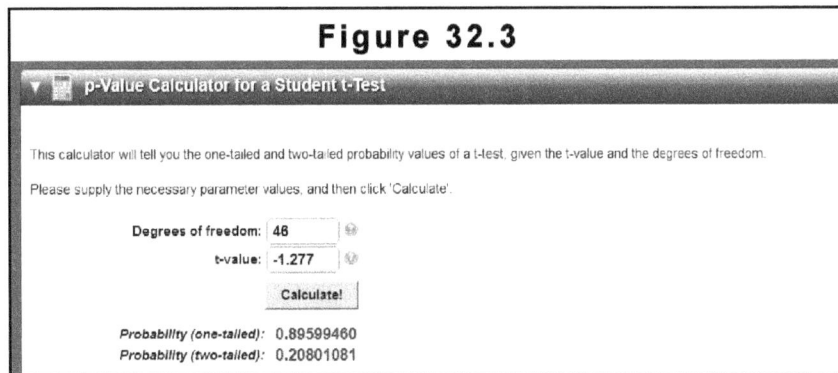

Figure 32.3

p-Value Calculator for a Student t-Test

This calculator will tell you the one-tailed and two-tailed probability values of a t-test, given the t-value and the degrees of freedom.

Please supply the necessary parameter values, and then click 'Calculate'.

Degrees of freedom: 46
t-value: -1.277
Calculate!

Probability (one-tailed): 0.89599460
Probability (two-tailed): 0.20801081

Endnote

[1] Soper, D. (2015). *p-value calculator.* URL:http://www.DanielSoper.com

Blank Page

Testing Two Paired Means

Chapter 31 described a t test for comparing a sample mean, a statistic, which is a proxy for a population mean, a parameter, to a hypothesized population mean. This test is known as a one-sample t test. And, Chapter 32 described a t test for comparing the means of two independent samples, which are proxies for two population means. This test is known as a two-sample t test. This chapter describes a t test for comparing the means of two paired samples. To start, review the methodologies described in Chapters 23 and 30.

Sample Data

SAT scores are approximately normally distributed. Twenty-five male students were tested before and after being tutored. Table 33.1 shows a random sample of the SAT scores of the students before tutoring, the SAT scores of the same students after tutoring, and the differences.

Table 33.1				
SAT Scores - Before				
504	519	462	444	530
347	538	561	709	463
640	427	450	538	469
410	521	520	494	392
369	431	487	361	506
SAT Scores - After				
519	509	476	435	546
354	538	572	709	472
646	431	455	543	474
410	531	520	504	392
362	444	477	372	496
Differences				
15	-10	14	-9	16
7	0	11	0	9
6	4	5	5	5
0	10	0	10	0
-7	13	-10	11	-10

Null Hypothesis and Alternative Hypothesis

Say that one states that the average SAT scores of students after tutoring is higher than the average SAT scores of students before tutoring. That is, H_0: $\mu_a = \mu_b$ and $\mu_a - \mu_b = 0$ and H_1: $\mu_a > \mu_b$. Is it true?

Figure 33.1 shows that, if the the test shows that the calculated t score is within the acceptance region, shown in gray, then, one fails to reject —that is, accepts—the null hypothesis and, therefore, is likely that there is no difference between the average SAT scores of students after tutoring and the average SAT scores of students before tutoring. Conversely, if the test shows that the the calculated t score is within the rejection region, shown in black, then, one rejects the null hypothesis and, therefore, is likely that the average SAT scores of students after tutoring is higher than the average SAT scores of students before tutoring.

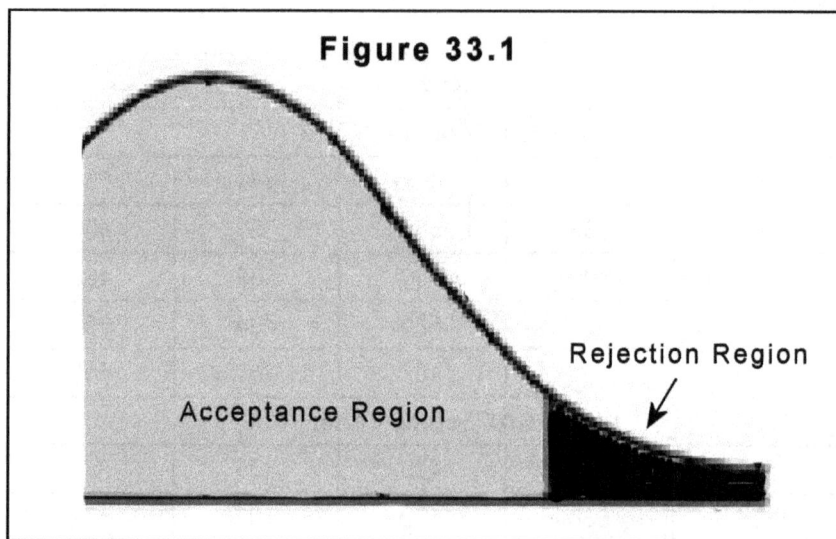

Figure 33.1

Acceptance Region

Rejection Region

Sample Mean

The sample mean of the distribution of the differences shown in Table 33.1 on Page 133 is 3.8:

$$(15+7+...+(-10))/25=3.8$$

Sample Standard Deviation

The sample standard deviation of the distribution of the differences shown in Table 33.1 on Page 133 is 8.20:

$$\text{SQRT}(((15\text{-}3.8)^2+(7\text{-}3.8)^2+...+((\text{-}10)\text{-}3.8)^2)/24)=8.20$$

T Score

The formula for calculating the t score is …

$$t=\dfrac{\overline{x}-(\mu_a-\mu_b)}{\dfrac{s}{\sqrt{n}}}$$

… where t is the t score, x-bar is the sample mean of the differences, μ_a and μ_b are the population means, s is the sample standard deviation of the differences, and n is the sample size.

For this example, $(\mu_a-\mu_b)=0$ and the relevant t score is 2.317:

$$(3.8\text{-}0)/(8.20/\text{SQRT}(25))=2.317$$

Degrees of Freedom

For this example, the degrees of freedom are 25-1=24.

Confidence Level and Critical Values

The t table is used for finding the critical values (CVs) of t that correspond to a selected level of confidence, where CL=100%-α, which, for this example, 97.5% is selected. Table 33.2 on Page 138 and Figure 33.2 on Page 138 show, for DF=24, that the t score that corresponds to the 97.5% confidence level, which is CL=100%-2.5%, is +2.064 for the right side. The confidence level shows the probability, which is 97.5%, that the t score is between -∞ and +2.064.

Table 33.2			
Degrees of Freedom	**Probability (One Tail)**		
	0.1000	0.0500	0.0250
	1.318	1.711	2.064
24	**Probability (Two Tail)**		
	0.1000 (0.0500+0.0500)	0.0500 (0.0250+0.0250)	0.0250 (0.0125+0.0125)
	1.711	2.064	2.391

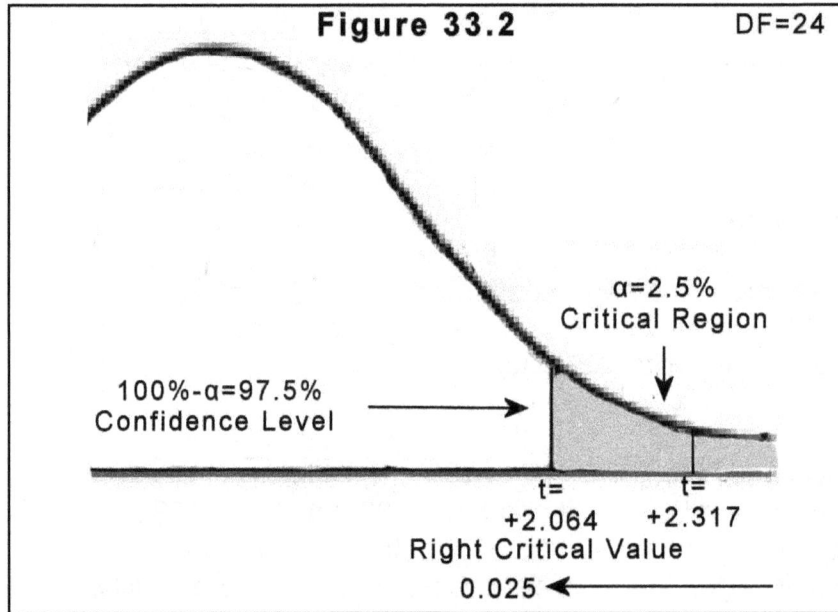

Figure 33.2   DF=24

α=2.5%
Critical Region

100%-α=97.5%
Confidence Level

t= +2.064   t= +2.317
Right Critical Value
0.025 ◄

Inference

Figure 33.2 shows that, for DF=24, the t score of +2.317 is more than the critical value of +2.064 and outside the 97.5% confidence level. Therefore, one rejects the null hypothesis. That is, one can say that, most likely, the average SAT scores of students after tutoring is higher than the average SAT scores of students before tutoring.

Z Test

Again review the methodology described in Chapter 23. If the standard deviation of the population is known, then one uses a z score rather than a t score. And, according to some, but not to all, if the sample size is greater than 30, then one also uses a z score rather than a t score.

P Value

As an alternative to the traditional test, one may select an alpha level, say $\alpha=2.5\%$, and calculate the p=value. If $p \leq \alpha$ one rejects H_0. For example, using a web-based calculator, for DF=24, the one-tail probability of a t score of +2.315 is 0.015 (See Figure 32.3 on Page 137). [1] And, p=0.015 is less than 2.5%. Thus, one rejects H_0.

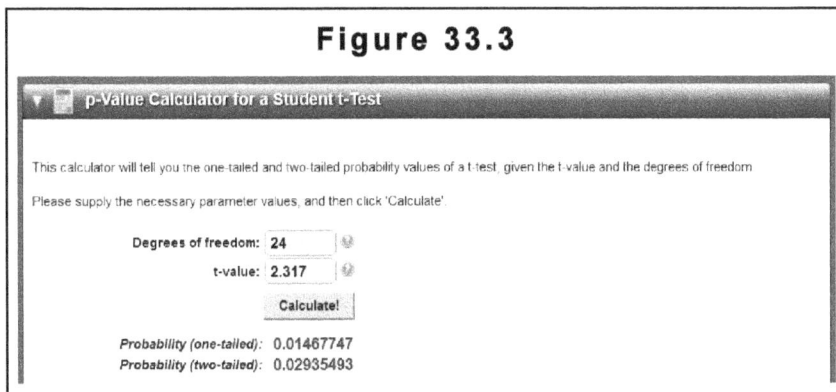

Figure 33.3

p-Value Calculator for a Student t-Test

This calculator will tell you the one-tailed and two-tailed probability values of a t-test, given the t-value and the degrees of freedom

Please supply the necessary parameter values, and then click 'Calculate'

Degrees of freedom: 24

t-value: 2.317

Calculate!

Probability (one-tailed): 0.01467747
Probability (two-tailed): 0.02935493

Endnote

[1] Soper, D. (2015). *p-value calculator*. URL:http://www.DanielSoper.com

Blank Page

Testing a Hypothesized Proportion

Recall that proportions describe the relationship between the frequencies of values and the whole. For example, about 2 out 5, or 2/5, or 0.40, or 40% of all adult Americans are college grads. This chapter describes a z test for comparing a sample proportion, a statistic, which is a proxy for a population proportion, a parameter, to a hypothesized population proportion. The ways for testing proportions parallel the ways of testing means.

Sample Data

In a randomized survey, 385 out of 1,000 adult Americans said that they graduated college.

Null Hypothesis and Alternative Hypothesis

Say that one states that 40% of all adult Americans graduated college. That is H_0: p=40% and H_1: p≠40%. Is it true?

Figure 34.1 shows that, if the the test shows that the calculated z score is within the acceptance region, shown in gray, then, one fails to reject —that is, accepts—the null hypothesis and, therefore, is likely that there is no difference between the percentage of adult Americans that graduated college and 40%. Conversely, if the test shows that the calculated z score is within the rejection regions, shown in black, then, one rejects the null hypothesis and, therefore, is likely that there is a difference between the percentage of adult Americans that graduated college and 40%.

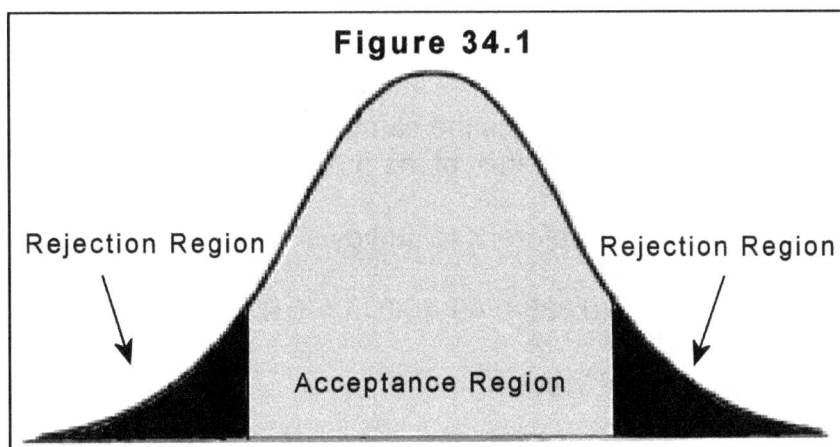

Figure 34.1

Rejection Region

Rejection Region

Acceptance Region

Sample Proportion

Recall that the sample proportion, p-hat, is the best estimate of the population proportion, p. The sample proportion for the above-referenced survey is 0.385:

$$385/1,000=0.385$$

Standard Error of Proportion

The formula for the standard error of the proportion, SE_p, is ...

$$SE_p = \sqrt{\frac{p*(1-p)}{n}}$$

... where SE_p is the standard error of the proportion, p is the proportion, and n is the sample size. Additionally, $np(1-p)>10$ and $n \leq 0.05N$.

The standard error of the proportion for the above-referenced survey is 0.015:

$$SQRT(0.40*(1-0.40)/1,000)=0.015$$

Z Score

The formula for calculating the z score is ...

$$Z = \frac{\text{p-hat} - p}{SE_p}$$

... where z is the z score, p-hat is the sample proportion, p is the hypothesized proportion, and SE_p is the standard error of the proportion.

The z score for the above-referenced survey is -0.96:

$$(0.385-0.40)/(0.015)=-0.96$$

Confidence Level and Critical Values

The z table is used for finding the critical values (CVs) of z that correspond to a selected level of confidence, where CL=100%-α, which, for this example, 90% is selected. Table 34.1 on Page 143 and Figure 34.2 on Page 143 show that the z scores that correspond to the 90% confidence level, which is CL=100%-10%, are -1.645 for the left side and +1.645 for the right side. The confidence level shows the probability, which is 90%, that the z scores are between -1.645 and +1.645. Note that $\alpha/2$=5.0%.

Table 34.1		
z Score		Cumulative Left-Tail Probability
-1.9	.60	0.02500
-1.6	.45	0.05000
-1.2	.82	0.10000
+1.2	.82	0.90000
+1.6	.45	0.95000
+1.9	.60	0.97500

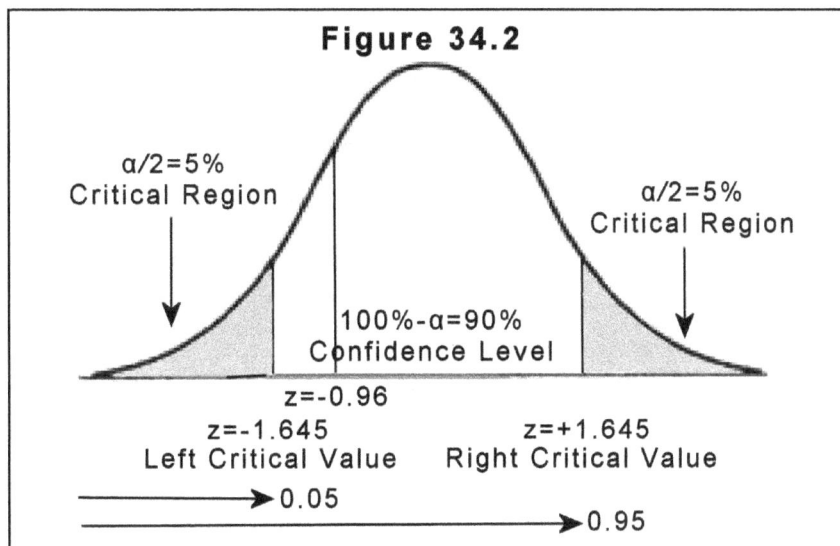

Figure 34.2

$\alpha/2$=5%
Critical Region

$\alpha/2$=5%
Critical Region

100%-α=90%
Confidence Level

z=-0.96
z=-1.645
Left Critical Value

z=+1.645
Right Critical Value

0.05

0.95

Inference

Figure 34.2 shows that the z score of -0.96 is between the critical values of -1.645 and +1.645 and within the 90% confidence level. Therefore, one does not reject the null hypothesis. That is, one can say that, most likely, the proportion of all adult Americans that graduated college is within the 90% confidence level.

P Value

As an alternative to the traditional test, one may select an alpha level, say α=10%, and calculate the p=value. If p>α one fails to reject —that is, accepts— H_0.

For example, using a web-based calculator the two-tail probability of a z score of -0.96 is 0.3335 (See Figure 34.3). [1] And, p=0.3345 is more than 10%. Thus, one does not reject H_0.

Figure 34.3
P Value from Z Score Calculator

This is very easy: just stick your Z score in the box marked Z score, select your significance level and whether you're testing a one or two-tailed hypothesis (if you're not sure, go with the defaults), then press the button!

If you need to derive a Z score from raw data, you can find a Z test calculator here.

Z score: -0.965

Significance Level:

○ 0.01
○ 0.05
◉ 0.10

One-tailed or two-tailed hypothesis?:

○ One-tailed
◉ Two-tailed

The P-Value is 0.334545.

The result is *not* significant at p < 0.10.

Calculate

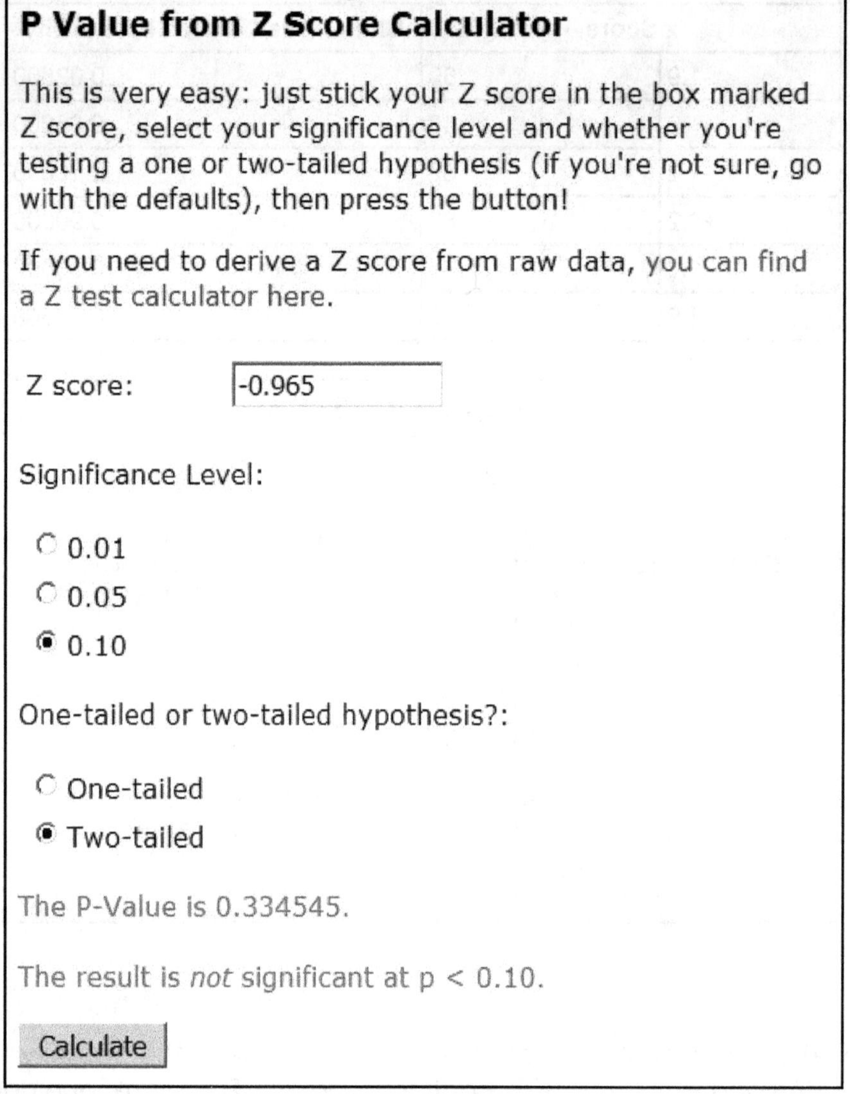

Endnote

[1] Social Science Statistics. (n.d.). *P value from z score calculator*. URL: http://www.socscistatistics.com/

Testing Two Independent Proportions

Chapter 34 described a z test for comparing a sample proportion, a statistic, which is a proxy for a population proportion, a parameter, to a hypothesized population proportion. This chapter describes a z test for comparing two independent proportions. Recall that the ways for testing proportions parallel the ways of testing means.

Sample Data

In a randomized survey, 670 out of 1,000 adult male Americans said that they support the death penalty but only 590 out of 1,000 adult female Americans said they do so.

Null Hypothesis and Alternative Hypothesis

Say that one states that, regarding support for the death penalty, there is no difference between males and females. That is, H_0: $p_m = p_f$ and H_1: $p_m \neq p_f$. Is it true?

Figure 35.1 shows that, if the the test shows that the calculated z score is within the acceptance region, shown in gray, then, one fails to reject —that is, accepts—the null hypothesis and, therefore, is likely that there is no difference between males and females. Conversely, if the test shows that the the calculated z score is within the rejection regions, shown in black, then, one rejects the null hypothesis and, therefore, is likely that there is a difference between males and females.

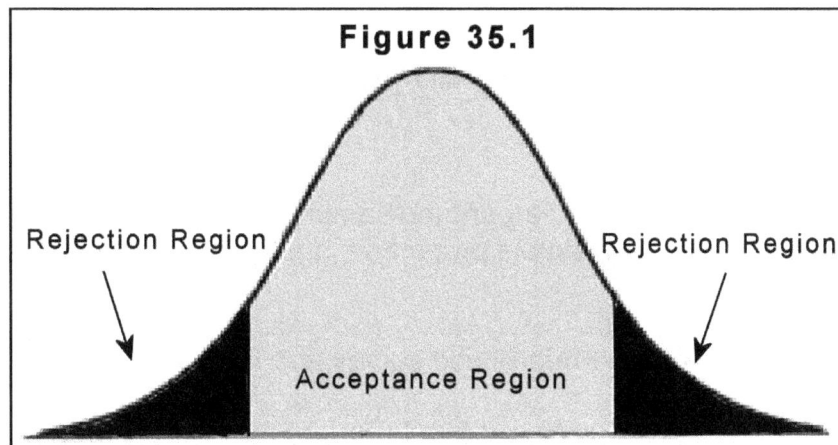

Figure 35.1

Rejection Region   Rejection Region

Acceptance Region

Sample Proportions

Recall that the sample proportion, p-hat, is the best estimate of the population proportion, p. The sample proportions for the above-referenced survey are 0.67 and

0.59:

$$670/1,000=0.67$$

$$590/1,000=0.59$$

Pooled Sample Proportion

The formula for the pooled sample proportion is …

$$\text{p-bar}=\frac{x_1+x_2}{n_1+n_2}$$

… where p-bar is the pooled sample proportion, x_s are the counts of the characteristics, n_s are the sample sizes, and p-hat$_s$>5 and (1 - p-hat$_s$)>5.

The pooled sample proportion for the above-referenced survey is 0.63:

$$(670+590)/(1,000+1,000)=0.63$$

Z Score

The formula for calculating the z score is …

$$z=\frac{(\text{p-hat}_1 - \text{p-hat}_2)-(p_1-p_2)}{\sqrt{\dfrac{\text{p-bar}(1 - \text{p-bar})}{n_1}+\dfrac{\text{p-bar}(1 - \text{p-bar})}{n_2}}}$$

… where z is the z score, p-hat$_s$ are the sample proportions, p_s are the proportions, p-bar is the pooled sample proportion, and n_s are the sample sizes. And, $p_1-p_2=0$.

The z score for the above-referenced survey is 3.71:

$$((0.67-0.59)-0)/\text{SQRT}((0.63*(1-0.63)/1,000)+(0.63*(1-0.63)/1,000))=3.71$$

Confidence Level and Critical Values

The z table is used for finding the critical values (CVs) of z that correspond to a selected level of confidence, where CL=100%-α, which, for this example, 95% is selected. Table 35.1 and Figure 35.2 show that the z scores that correspond to the 95% confidence level, which is CL=100%-5%, are -1.96 for the left side and +1.96 for the right side. The confidence level shows the probability, which is 95%, that the z scores are between -1.96 and +1.96. Note that $\alpha/2$=2.5%.

Table 35.1		
z Score		Cumulative Left-Tail Probability
-1.9	.60	0.02500
-1.6	.45	0.05000
-1.2	.82	0.10000
+1.2	.82	0.90000
+1.6	.45	0.95000
+1.9	.60	0.97500

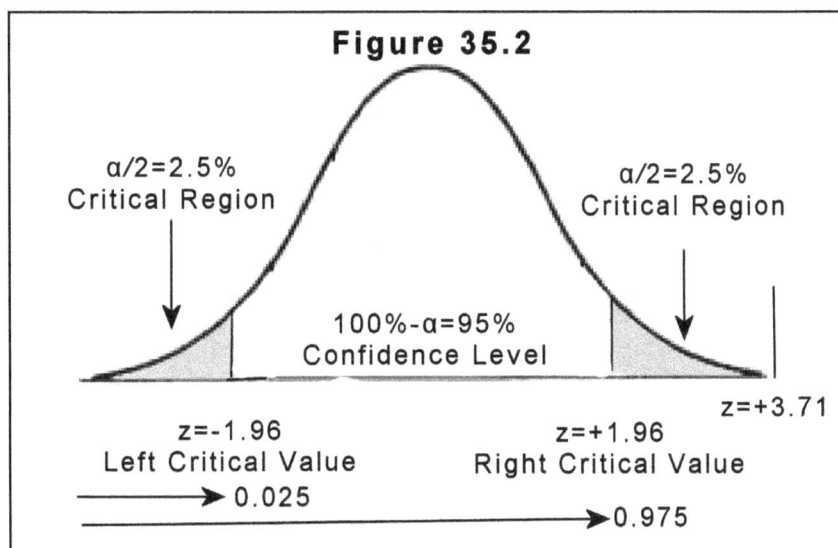

Figure 35.2

$\alpha/2$=2.5%
Critical Region

$\alpha/2$=2.5%
Critical Region

100%-α=95%
Confidence Level

z=+3.71

z=-1.96
Left Critical Value

z=+1.96
Right Critical Value

0.025

0.975

Inference

Figure 35.2 shows that the z score of +3.71 is way outside the critical values of -1.96 and +1.965 and way outside the 95% confidence level. Therefore, one rejects the null hypothesis. That is, one can say that, most likely, there is a difference between males and females.

P Value

As an alternative to the traditional test, one may select an alpha level, say α=5%, and calculate the p=value. If p<α one rejects H_0. For example, using a web-based calculator the two-tail probability of a z score of +3.71 is 0.0002 (See Figure 35.3). [1] And, p=0.0002 is less than than 5%. Thus, one rejects H_0.

Figure 35.3
P Value from Z Score Calculator

This is very easy: just stick your Z score in the box marked Z score, select your significance level and whether you're testing a one or two-tailed hypothesis (if you're not sure, go with the defaults), then press the button!

If you need to derive a Z score from raw data, you can find a Z test calculator here.

Z score: 3.71

Significance Level:

○ 0.01
◉ 0.05
○ 0.10

One-tailed or two-tailed hypothesis?:

○ One-tailed
◉ Two-tailed

The P-Value is 0.000207.

The result is significant at p < 0.05.

Calculate

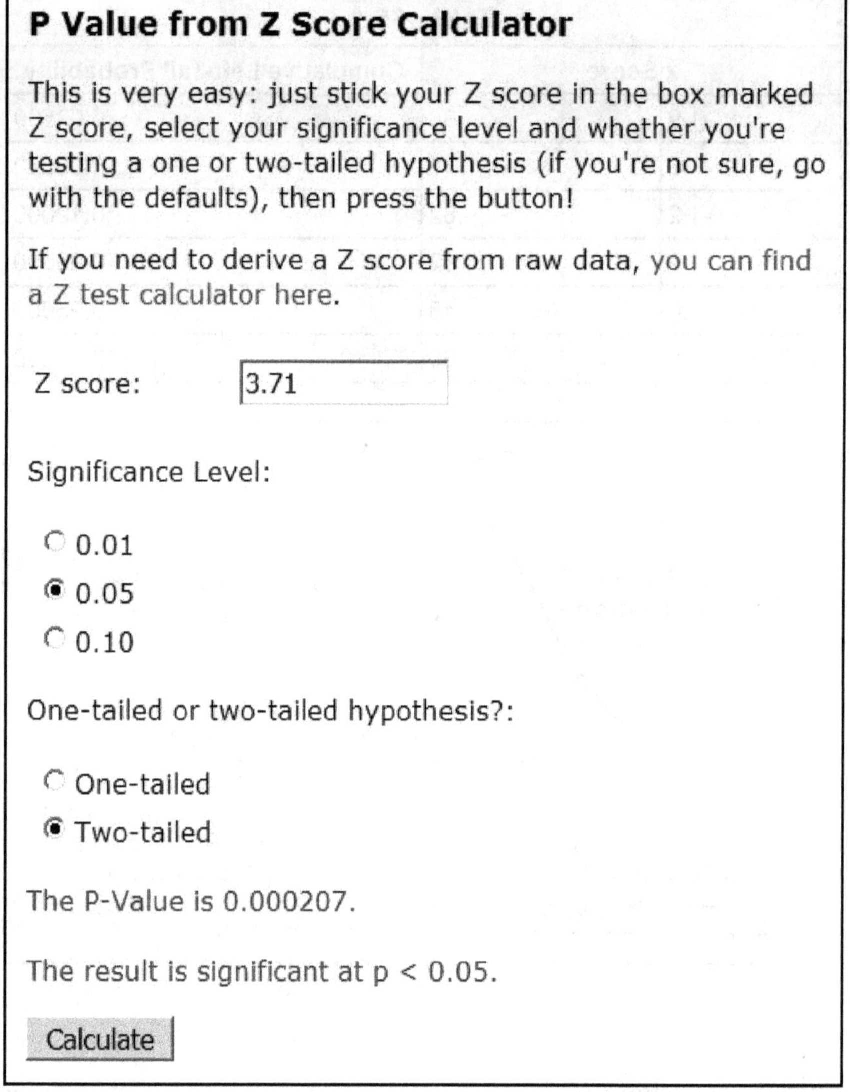

Endnote

[1] Social Science Statistics. (n.d.). *P value from z score calculator*. URL: http://www.socscistatistics.com/

Testing a Hypothesized Variance

Recall that variances and standard deviations are measures of variability. The ways of testing variances parallel the ways of testing means and proportions. However, means and proportions follow a normal distribution whereas variances follow a chi-square distribution. Therefore, the left-side and the right-side critical values are not equal (See Figure 36.1). This chapter describes a chi-square test for comparing a sample variance, a statistic, which is a proxy for a population variance, a parameter, to a hypothesized population variance.

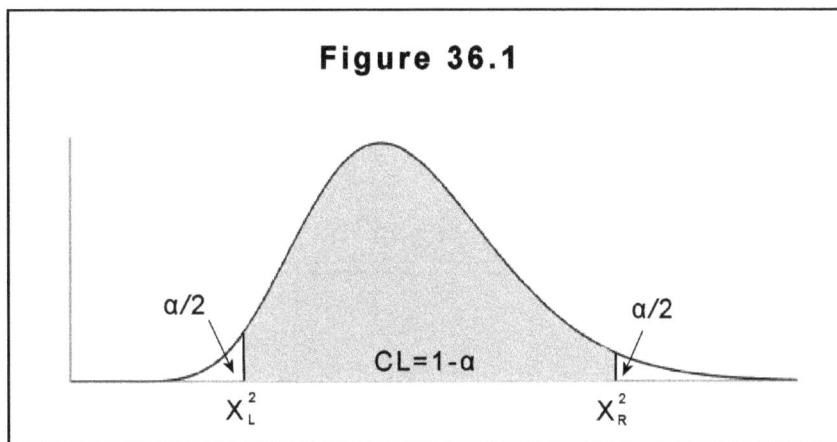

Figure 36.1

Sample Data

Table 36.1 shows the weights of a random sample of 25 18 oz. boxes of Kellogg's Corn Flakes.

Table 36.1				
17.77218	17.77069	18.19679	18.44214	18.10034
17.35447	17.97163	18.17372	18.01767	17.72556
18.71982	18.57064	18.04066	18.11353	17.99455
18.32668	17.93287	17.45444	18.03940	17.82327
17.75590	17.77189	18.34729	18.44081	17.34614

Null Hypothesis and Alternative Hypothesis

Say that one states that the standard deviation is ¼ oz. That is H_0: $\sigma = ¼$ oz. and H_1: $\sigma \neq ¼$. Is it true?

Figure 36.2 on Page 150 shows that, if the the test shows that the calculated chi-square statistic is within the acceptance region, shown in gray, then, one fails to reject —that is, accepts— the null hypothesis and, therefore, is likely that the standard deviation is about ¼ oz. Conversely, if the test shows that the calculated chi-square statistic is within the rejection regions, shown in black, then, one rejects the null hypothesis and, therefore, is likely that the standard deviation is not about ¼ oz.

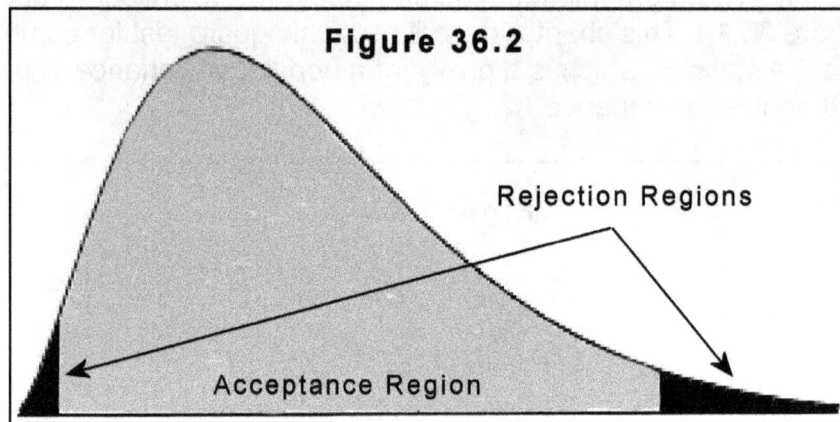

Figure 36.2

Rejection Regions

Acceptance Region

Sample Mean

The sample mean of the distribution shown in Table 36.1 on Page 149 is 18.01:

$$(17.77218+17.35447 +...+17.34614)/25=18.01$$

Sample Variance

The sample variance of the distribution shown in Table 36.1 on Page 149 is 0.126:

$$((17.77218-18.01)^2+(17.35447-18.01)^2+...+(17.34614-18.01)^2)/24=0.126$$

And, the sample standard deviation of the distribution shown in Table 36.1 on Page 149 is 0.355:

$$SQRT(0.126)=0.355$$

Chi-Square Statistic

The formula for calculating the chi-square statistic is ...

$$X^2 = \frac{(n-1)*s^2}{\sigma^2}$$

… where X^2 is the chi-square statistic, n is the sample size, s^2 is the sample variance, and σ^2 is the population variance.

The chi-square statistic of the distribution sown in Table 36.1 on Page 149 is 48.384:

$$(25-1)*0.126/0.25^2 = 48.384$$

Degrees of Freedom

For this example, the degrees of freedom are 25-1=24.

Confidence Level and Critical Values

The chi-square table is used for finding the critical values (CVs) of chi-square that correspond to a selected level of confidence, where CL=100%-α, which, for this example, 99% is selected. Table 36.2 and Figure 36.3 on Page 152 show that the chi-square statistics that correspond to the 99% confidence level, which is CL=100%-1%, are +9.886 for the left side and +45.559 for the right side. The confidence level shows the probability, which is 99%, that the chi-square statistics are between +9.886 and +45.559. Note that $\alpha/2$=0.005%.

Table 36.2				
Degrees of Freedom	Probability (Right-Tail)			
	0.995	0.900	0.10	0.005
24	9.886	15.659	33.196	45.559

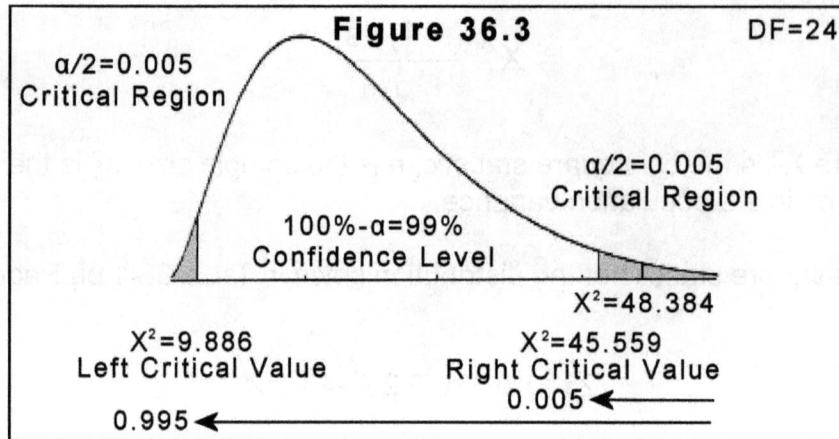

Figure 36.3 DF=24

$\alpha/2$=0.005
Critical Region

$\alpha/2$=0.005
Critical Region

100%-α=99%
Confidence Level

X^2=48.384

X^2=9.886
Left Critical Value

X^2=45.559
Right Critical Value
0.005

0.995

Inference

Figure 36.3 shows that the chi-square statistic of +48.384 is to the right of the critical value of +45.559 and outside the 99% confidence level. Therefore, one rejects the null hypothesis. That is, one can say that, most likely, the standard deviation is not about ¼ oz.

P Value

As an alternative to the traditional test, one may select an alpha level, say α=1%, and calculate the p=value. If p<α one rejects H_0. For example, using a web-based calculator, the probability of a chi-square statistic of 48.384 with DF=24 is 0.0022 (See Figure 36.4 on Page 153). [1] And, p=0.0022 is less than 1%. Thus, one rejects H_0.

Figure 36.4

P Value from Chi-square Calculator

This should be self-explanatory, but just in case it's not: your chi-square score goes in the chi-square score box, you stick your degrees of freedom in the *DF* box (*df = ($N_{Columns}$-1)*(N_{Rows}-1)* for chi-square test for independence), select your significance level, then press the button.

If you need to derive a chi-square score from raw data, then you can find chi-square test calculators here.

Chi-square score: 48.384

DF: 24

Significance Level:

⦿ 0.01

◯ 0.05

◯ 0.10

The P-Value is 0.002262. The result is significant at p < 0.01.

Calculate

Endnote

[1] Social Science Statistics. (n.d.). *P value from chi-square calculator.* URL: http://www.socscistatistics.com/

Blank Page

Testing Two Independent Variances

Chapter 36 described a chi-square test for comparing a sample variance, a statistic, which is a proxy for a population variance, a parameter, to a hypothesized population variance. This chapter describes an test for comparing two independent variances. This test is known as an f test. F tests are based on f distributions (See Figure 37.1).

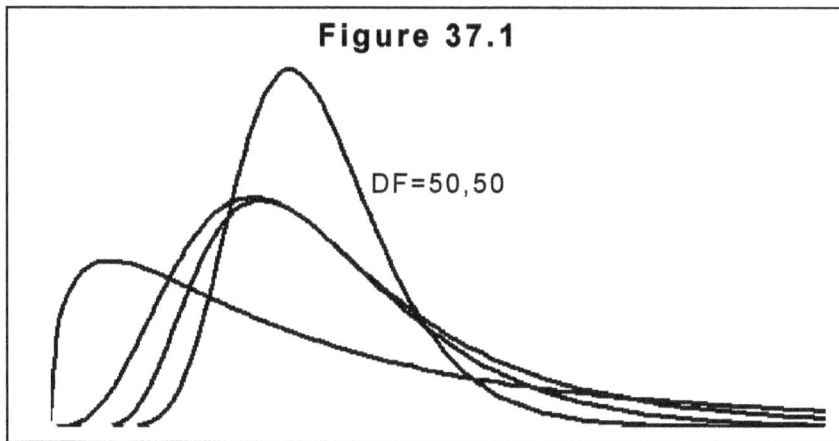

Figure 37.1

DF=50,50

Sample Data

Table 37.1 shows the weights of two random samples: 25 18 oz. boxes of Kellogg's Corn Flakes and 25 18 oz. boxes of Kellogg's Rice Krispies.

Table 37.1				
Corn Flakes				
17.77218	17.77069	18.19679	18.44214	18.10034
17.35447	17.97163	18.17372	18.01767	17.72556
18.71982	18.57064	18.04066	18.11353	17.99455
18.32668	17.93287	17.45444	18.03940	17.82327
17.75590	17.77189	18.34729	18.44081	17.34614
Rice Krispies				
18.42597	17.79911	17.47209	17.76121	17.59736
18.47256	17.87146	18.12279	18.28046	18.27157
17.55510	18.48723	17.83292	17.88519	18.20168
17.77491	17.91214	17.63455	18.69106	17.68452
17.91889	18.42287	17.92778	18.54321	18.08939

Null Hypothesis and Alternative Hypothesis

Say that one states that the standard deviations are equal That is H_0: $\sigma_{CF} = \sigma_{RK}$. and H_1: $\sigma_{CF} \neq \sigma_{RK}$. Is it true?

Figure 37.2 shows that, if the the test shows that the calculated f statistic is within the acceptance region, shown in gray, then, one fails to reject —that is, accepts— the null hypothesis and, therefore, is likely that the standard deviations are equal. Conversely, if the test shows that the calculated f statistic is within the rejection regions, shown in black, then, one rejects the null hypothesis and, therefore, is likely that the standard deviations are not equal.

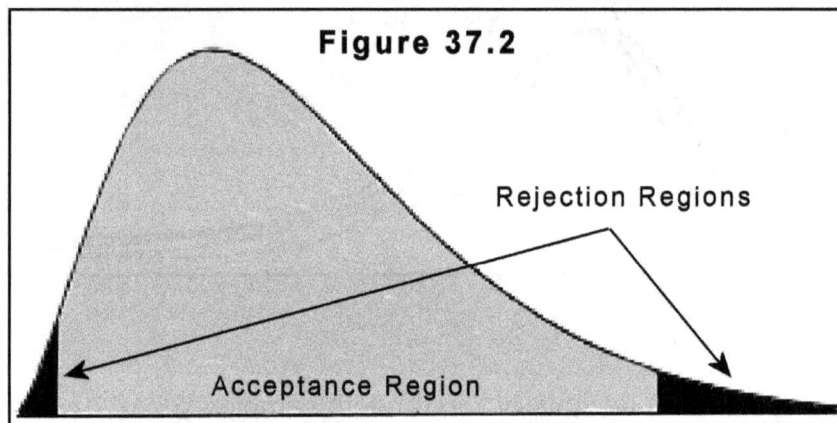

Figure 37.2

Rejection Regions

Acceptance Region

Sample Means

The sample means of the distributions sown in Table 36.1 on Page 155 are 18.01 and 18.03:

$$(17.77218 + 17.35447 + ... + 17.34614)/25 = 18.01$$

$$(18.42597 + 18.47256 + ... + 18.08939)/25 = 18.03$$

Sample Variances

The sample variances of the distributions sown in Table 36.1 on Page 155 are 0.126 and 0.120:

$$((17.77218-18.01)^2 + (17.35447-18.01)^2 + ... + (17.34614-18.01)^2)/24 = 0.126$$

$$((18.42597-18.03)^2 + (18.47256-18.03)^2 + ... + (18.08939-18.03)^2)/24 = 0.120$$

And, the sample standard deviations of the distributions sown in Table 36.1 on Page 155 are 0.355 and 0.347:

$$SQRT(0.126)=0.355$$

$$SQRT(0.120)=0.347$$

F Statistic

The formula for calculating the f statistic is …

$$f = \frac{s_{CF}^2 \ (\text{Larger variance})}{s_{RK}^2 \ (\text{Smaller variance})}$$

… where f is the f statistic and s_{CF}^2 and s_{RK}^2 are the sample variances.

The relevant f statistic is 1.05:

$$0.126/0.120=1.05$$

Degrees of Freedom

Te relevant degrees of freedom are 25-1=24 and 25-1=24.

Confidence Level and Critical Values

F tables are used for finding the critical values (CVs) of f that correspond to a selected level of confidence, where CL=100%-α, which, for this example, 95% is selected. Table 37.2 on Page 158 and Figure 37.3 on Page 158 show that, for DF=24,24, the f statistics that correspond to the 95% confidence level, which is CL=100%-5%, is 2.2693 for the right side. Note that α/2=2.5%. The left side critical value is found as follows:

- Flip the degrees of freedom and find the f statistic for α/2=2.5% and DF=24,24, which is 2.2693.
- Find the reciprocal of 2.2693, which is 1/2.2693 or 0.4419.

The confidence level shows the probability, which is 0.95%, that the f statistics are between 0.4419 and 2.2693. Note that α/2=2.5%.

Table 37.2			
α=2.5%			
	DF_1=1	DF_1=24	DF_1=∞
DF_2=1	647.7890	997.2492	1,018.258
DF_2=24	5.7166	2.2693	1.935
DF_2=∞	5.0239	1.6402	1.000

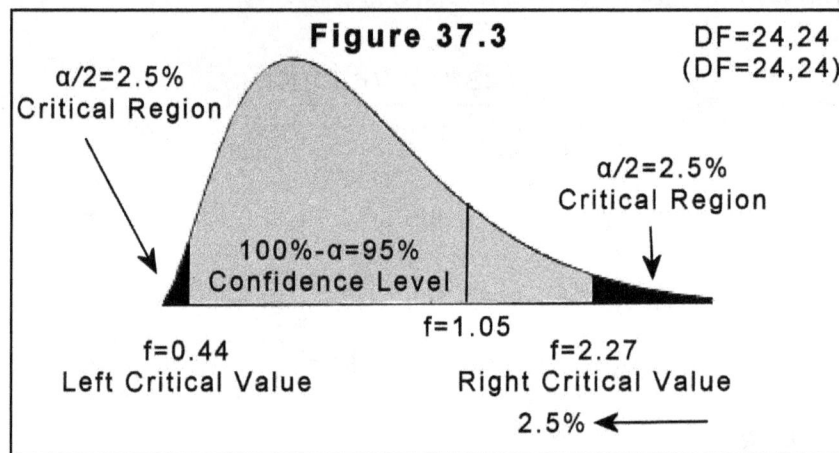

Figure 37.3   DF=24,24 (DF=24,24)

$\alpha/2$=2.5% Critical Region

$\alpha/2$=2.5% Critical Region

100%-α=95% Confidence Level

f=1.05

f=0.44 Left Critical Value

f=2.27 Right Critical Value

2.5%

In practice, only the right critical value is found. As shown above, the f statistic is always greater than one and the inverse is always less than one.

Inference

Figure 37.3 shows that the f statistic of 1.05 is between the critical values of 0.44 and 2.27 and within the 95% confidence level. Therefore, one does not reject the null hypothesis. That is, one can say that, most likely, there is no difference between the standard deviations.

P Value

As an alternative to the traditional test, one may select an alpha level, say α=5%, and calculate the p=value. If p>α one does not reject H_0. For example, using a web-based calculator, the probability of an f statistic of 1.05 with DF-24,24 is 0.45 (See Figure 37.4 on Page 159). [1] And, p=0.45 is more than 5%. Thus, one does not reject H_0.

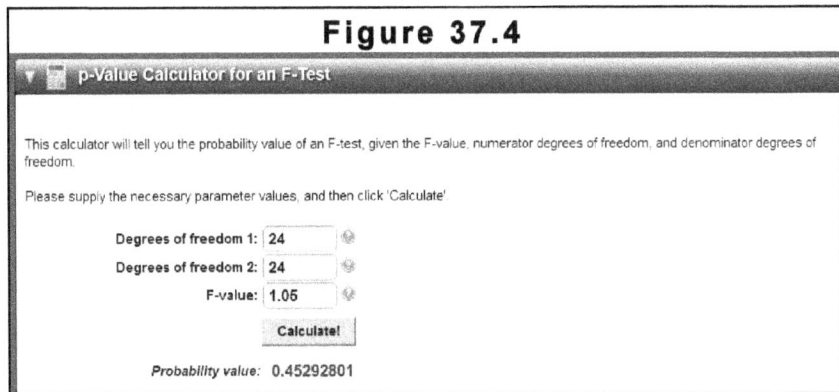

Figure 37.4

▼ p-Value Calculator for an F-Test

This calculator will tell you the probability value of an F-test, given the F-value, numerator degrees of freedom, and denominator degrees of freedom.

Please supply the necessary parameter values, and then click 'Calculate'

Degrees of freedom 1: 24

Degrees of freedom 2: 24

F-value: 1.05

Calculate!

Probability value: 0.45292801

Endnote

[1] Soper, D. (2015). *p-value calculator for an f test*. URL:http://www.DanielSoper.com

Blank Page

Goodness-of-Fit Tests

Goodness-of-fit tests are used to compare the observed frequencies, O, of one categorical variables versus the expected frequencies, E, when $n_E > 5$. For example, goodness-of-fit tests are used to find out if the colors of M&M Milk Chocolates are evenly distributed. Goodness-of-fit tests use chi-square statistics.

Sample Data

Table 38.1 shows the colors of a 1.69 oz. bag of 60 M&M Milk Chocolates.

Table 38.1

Blue	Brown	Green	Orange	Red	Yellow
17	6	6	15	10	6

Null Hypothesis and Alternative Hypothesis

Say that one states that the colors of M&M Milk Chocolates are evenly distributed; that their frequencies are equal. That is, $H_0: O_1 = O_2 = O_3 = O_4 = O_5 = O_6$ and the alternative hypothesis, H_1, is that at least two of the frequencies are not equal. Is it true?

Figure 38.1 on Page 162 shows that, if the the test shows that the calculated chi-square statistic is within the acceptance region, shown in gray, then, one fails to reject —that is, accepts— the null hypothesis and, therefore, is likely that the colors of M&M Milk Chocolates are evenly distributed. Conversely, if the test shows that the calculated chi-square statistic is within the rejection region, shown in black, then, one rejects the null hypothesis and, therefore, is likely that the colors of M&M Milk Chocolates are not evenly distributed.

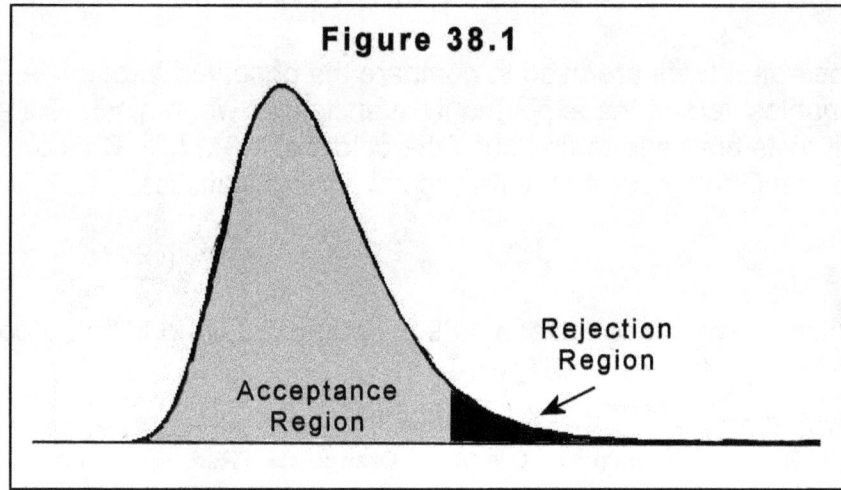

Figure 38.1

Chi-Square

The formula for calculating the chi-square statistic is ...

$$X^2 = \sum \frac{(O-E)^2}{E}$$

... where X^2 is the chi-square statistic, O are the observed frequencies, and E are the expected frequencies.

The relevant chi-square statistic is 12.2:

$$(17-10)^2/10 + (6-10)^2/10 + ... + (6-10)^2/10 = 12.2$$

Degrees of Freedom

Te relevant degrees of freedom are 6-1=5.

Confidence Level and Critical Values

Chi-square tables are used for finding the critical values (CVs) of chi-square that correspond to a selected level of confidence, where CL=100%-α, which, for this example, 95% is selected. Table 38.2 on Page 163 and Figure 38.2 on Page 163 show, for DF=24, that the chi-square statistic that correspond to the 95% confidence level, which is CL=100%-5%, is 11.070 for the right side. The confidence level shows the probability, which is 95%, that the chi-square statistic is less than 11.070.

Table 38.2				
Degrees of Freedom	**Probability (Right-Tail)**			
	0.100	0.050	0.025	0.001
24	9.236	11.070	12.833	15.086

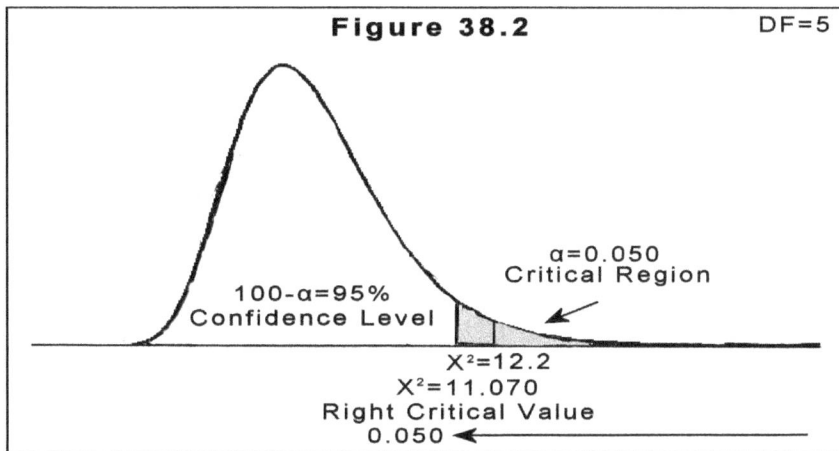

Figure 38.2   DF=5

α=0.050
Critical Region

100-α=95%
Confidence Level

X^2=12.2
X^2=11.070
Right Critical Value
0.050

Inference

Figure 38.2 shows that the chi-square statistic of 12.2 is to the right of the critical value of 11.070 and outside the 95% confidence level. Therefore, one rejects the null hypothesis. That is, one can say that, most likely, there is a difference between at least two of the frequencies.

P Value

As an alternative to the traditional test, one may select an alpha level, say α=5%, and calculate the p=value. If p<α one rejects H_0. For example, using a web-based calculator, the probability of a chi-square statistic of 12.2 with DF=5 is 0.032 (See Figure 38.3 on Page 162). [1] And, p=0.032 is less than 5%. Thus, one rejects H_0.

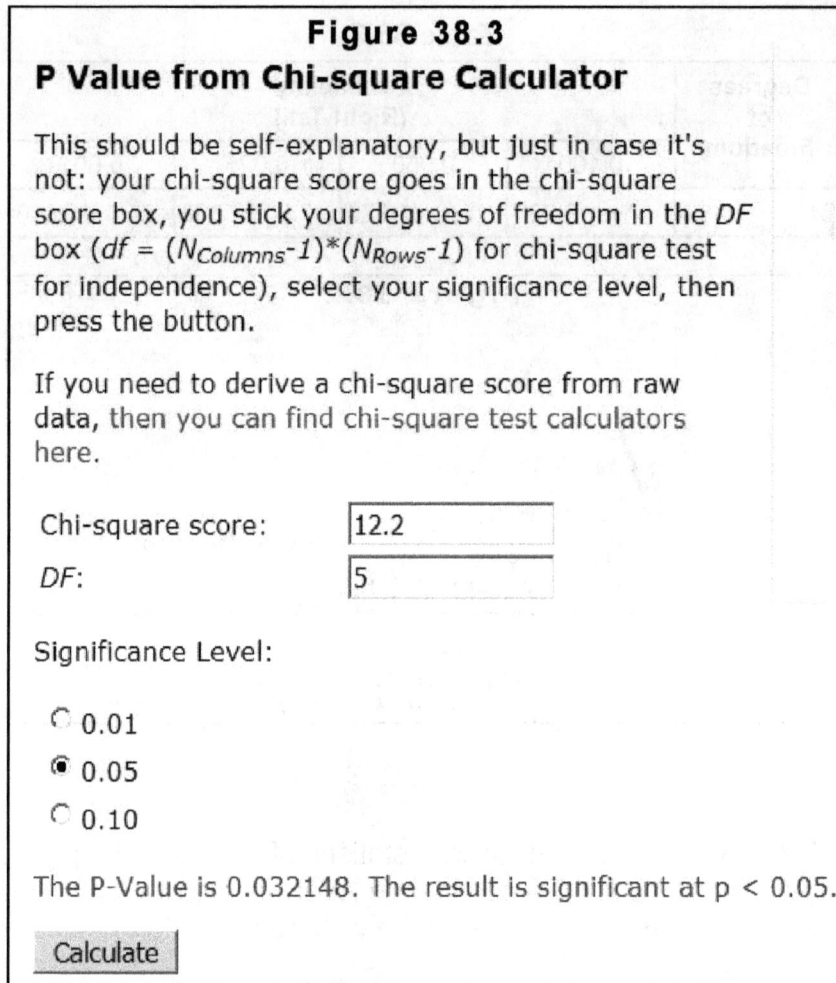

Figure 38.3
P Value from Chi-square Calculator

This should be self-explanatory, but just in case it's not: your chi-square score goes in the chi-square score box, you stick your degrees of freedom in the *DF* box ($df = (N_{Columns}-1)*(N_{Rows}-1)$ for chi-square test for independence), select your significance level, then press the button.

If you need to derive a chi-square score from raw data, then you can find chi-square test calculators here.

Chi-square score: `12.2`

DF: `5`

Significance Level:

○ 0.01

◉ 0.05

○ 0.10

The P-Value is 0.032148. The result is significant at $p < 0.05$.

Calculate

Endnote

[1] Social Science Statistics. (n.d.). *P value from chi-square calculator*. URL: http://www.socscistatistics.com/

Contingency Tables

Contingency tables show the frequencies of two categorical variables. For example, Table 39.1 shows the observed frequencies of cell-phone users between males and females. Chi-square tests are used to compare the observed frequencies to the expected frequencies of the two variables.

Sample Data

Table 39.1 shows the observed frequencies of cell-phone users of a random sample of 500 male and 500 female Americans.

Table 39.1			
Observed			
	Male	Female	Total
Users	465	440	905
Non-Users	35	60	95
Total	500	500	1,000

The row totals and the column totals are labeled marginal totals and the total of all the frequencies is labeled the grand total.

Observed Frequencies versus Expected Frequencies

Based on observed frequencies, one may built a contingency table of expected frequencies. The formula for calculating the expected frequencies is ...

$$EF = \frac{RT \sum CT}{GT}$$

... where EF is the expected frequency, RT is the row total, CT is the column total, and GT is the grand total. For example, the expected frequency of male users is 452.5:

$$(905*500)/1,000 = 452.5$$

Table 39.2 on Page 166 shows the expected frequencies of cell-phone users of the random sample of 500 male and 500 female Americans shown in Table 39.1.

Table 39.2			
Expected			
	Male	**Female**	**Total**
Users	452.5	452.5	905
Non-Users	47.5	47.5	95
Total	500	500	1,000

Null Hypothesis and Alternative Hypothesis

Say that one states that the frequencies of male users is equal to the frequencies of the female users. That is, $H_0: O_m = O_f$ and $H_1: O_m \neq O_f$. Is it true?

Figure 39.1 shows that, if the the test shows that the calculated chi-square statistic is within the acceptance region, shown in gray, then, one fails to reject —that is, accepts— the null hypothesis and, therefore, is likely that the frequencies of male users is equal to the frequencies of the female users. Conversely, if the test shows that the calculated chi-square statistic is within the rejection region, shown in black, then, one rejects the null hypothesis and, therefore, is likely that the frequencies of male users is not equal to the frequencies of the female users.

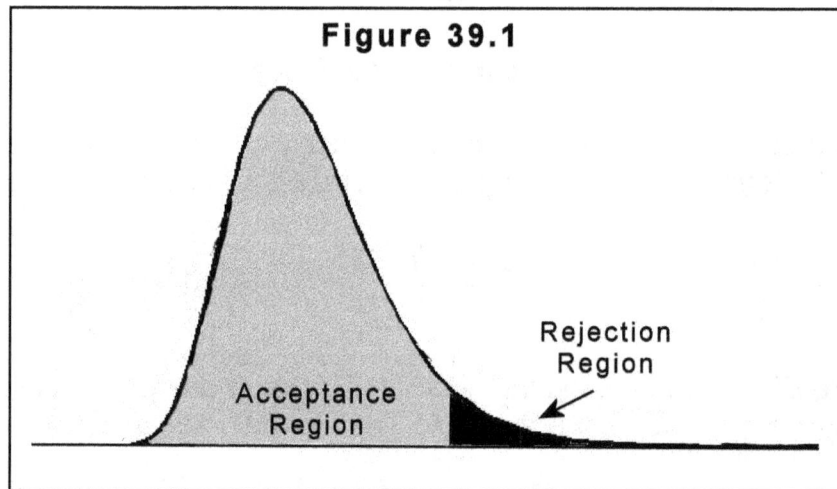

Figure 39.1

Chi-Square

The formula for calculating the chi-square statistic is ...

$$X^2 = * \frac{(O-E)^2}{E}$$

... where X^2 is the chi-square statistic, O are the observed frequencies, and E are the expected frequencies.

The relevant chi-square statistic is 7.27:

$$(465-452.5)^2/452.5+(440-452.5)^2/452.5+...+(60-47.5)^2/47.5=7.27$$

Degrees of Freedom

The formula for calculating the degrees of freedom is ...

$$DF=(R-1)\Sigma(C-1)$$

... where DF is the degrees of freedom, R is the number of rows, and C is the number of columns.

The relevant degrees of freedom is 1:

$$(2-1)*(2-1)=1$$

Confidence Level and Critical Values

Chi-square tables are used for finding the critical values (CVs) of chi-square that correspond to a selected level of confidence, where CL=100%-α, which, for this example, 99% is selected. Table 39.3 and Figure 39.2 on Page 168 show that, for DF=1, the chi-square statistic that correspond to the 99% confidence level, which is CL=100%-1%, is 6.635 for the right side. The confidence level shows the probability, which is 99%, that the chi-square statistic is less than 6.635.

Table 39.3				
Degrees of Freedom	**Probability (Right-Tail)**			
	0.100	0.050	0.025	0.010
1	2.706	3.841	5.024	6.635

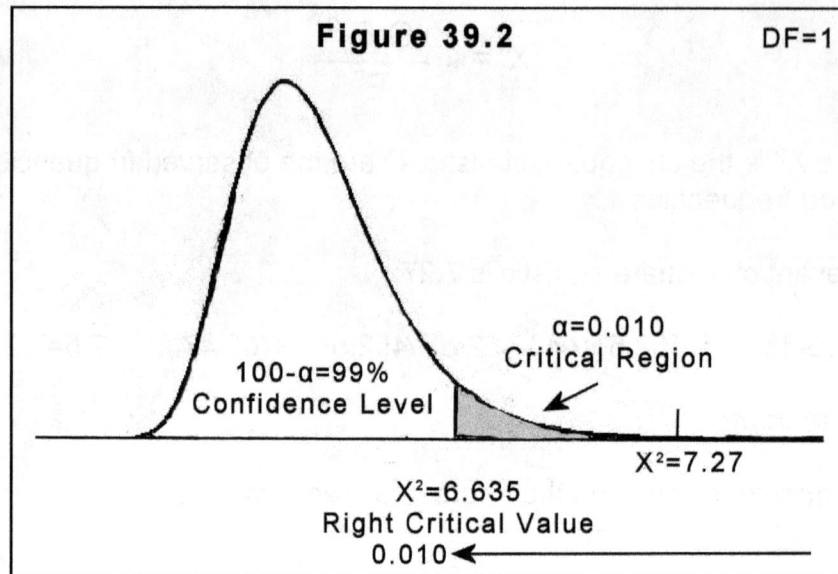

Figure 39.2   DF=1

α=0.010 Critical Region

100-α=99% Confidence Level

X²=7.27

X²=6.635 Right Critical Value

0.010

Inference

Figure 39.2 shows that the chi-square statistic of 7.27 is to the right of the critical value of 6.635 and outside the 99% confidence level. Therefore, one rejects the null hypothesis. That is, one can say that, most likely, the frequencies of male users is not equal to the frequencies of the female users.

P Value

As an alternative to the traditional test, one may select an alpha level, say α=1%, and calculate the p=value. If p<α one rejects H_0. For example, using a web-based calculator, the two-tail probability of a chi-square statistic of 7.27 with DF=1 is 0.007 (See Figure 39.3 on Page 169). [1] And, p=0.007 is less than 1%. Thus, one rejects H_0.

Figure 39.3

P Value from Chi-square Calculator

This should be self-explanatory, but just in case it's not: your chi-square score goes in the chi-square score box, you stick your degrees of freedom in the *DF* box (*df* = $(N_{Columns}-1)*(N_{Rows}-1)$ for chi-square test for independence), select your significance level, then press the button.

If you need to derive a chi-square score from raw data, then you can find chi-square test calculators here.

Chi-square score: 7.27

DF: 1

Significance Level:

⦿ 0.01

◌ 0.05

◌ 0.10

The P-Value is 0.007012. The result is significant at p < 0.01.

Calculate

Other Tests

This chapter describes the chi-square test without the Yates correction for continuity, which subtracts 0.5 from the difference between each observed value and its expected value in a 2×2 contingency table. [2] As an alternative to the chi-square test, one might use the Fisher exact test.

Endnotes

[1] Social Science Statistics. (n.d.). *P value from chi-square calculator*. URL: http://www.socscistatistics.com/

[2] Yates, F. (1934). Contingency table involving small numbers and the χ^2 test.

Supplement to the Journal of the Royal Statistical Society, 1(2), 217–235.

Introduction to ANOVA

In the analysis of variance, ANOVA, one compares a characteristic for three or more groups. Furthermore, in ANOVA one uses variances, which are the deviations of the values from the means. Finally, in ANOVA one uses the f statistic, which, in ANOVA, is the ratio of the explained variance to the unexplained variance.

Types of ANOVA

The design of ANOVA is contingent on the data. For example, in a one-way ANOVA one compares a characteristic for one variable, called a factor. And, in a two-way ANOVA one compares a characteristic for two factors.

One-Way ANOVA. In a one-way ANOVA one compares a characteristic for one factor for three or more groups. For example, Table 40.1 shows the SAT math scores for five male students from three high schools where the factor is the school.

Table 40.1 (Male Students)		
School X	**School Y**	**School Z**
521	647	506
365	475	475
644	631	318
562	643	436
542	788	350

Two-Way ANOVA Without Replication. In a two-way ANOVA without replication one compares a characteristic for two factors when there is only one value for each combination of factors. For example, Table 40.2 shows the SAT math scores for three male students for three tests where the factors are the student and the test.

Table 40.2			
	Test # 1	**Test # 2**	**Test # 3**
Student # 1	580	595	539
Student # 2	625	718	736
Student # 3	558	435	417

Two-Way ANOVA With Replication. In a two-way ANOVA with replication one compares a characteristic for two factors when there is more than one value for each combination of factors. For example, Table 40.3 shows the SAT math scores for five male students and five female students from three high schools where the factors are the gender and the school.

Table 40.3			
	School X	**School Y**	**School Z**
Male Students	521	647	506
	365	475	475
	644	631	318
	562	643	436
	542	788	350
Female Students	294	698	333
	566	431	265
	418	483	358
	428	577	527
	457	606	194

Variances

Recall that in ANOVA one measures variances. Also recall that variances are the deviations of the values from the mean.

Deviate. In a group of values, a deviate is the difference between a value and the group mean.

Squared Deviate. A squared deviate is, obviously, the square of a deviate.

Sum of Squares. The sum of the squared deviates, SS, is, also obviously, the sum of all of the squared deviates.

Mean Square. The mean of the squared deviates, MS —the average variance— is equal to the sum of squares, SS, divided by the appropriate degrees of freedom, DF.

Sources of Variances

In ANOVA, the variances are due to (1) the effect of the factors or (2) the effect of randomness. Table 40.3 on Page 173 shows the sources of the variances.

Table 40.3			
	One-Way ANOVA		Two-Way ANOVA
Factor (Between Groups)	X	Factor A	X
		Factor B	X
		Factor A x B	X
Randomness (Within Groups)	X		X
Total	X		X

Formulas

This section describes the formulas used in a one-way ANOVA.

Group Means. The group means are equal to the sum of the values of the group divided by the sample size of the group. The formula for the group mean is ...

$$\overline{x}_G = \frac{\sum x_G}{n_G}$$

... where x-bar$_G$ is the group mean, x_G are the group values, and n_G is the sample size of the group.

Grand Mean. The grand mean is equal to the sum of all the values divided by the total number of values. The formula for the grand mean is ...

$$\overline{x}_T = \frac{\sum x}{n_T}$$

... where x-bar$_T$ is the grand mean, x are the values, and n_T is the total number of values.

Sum of Squares Between Groups. The sum of the squared deviates between groups, SS_{BG}, is equal to the sum of the weighted differences between the group means and the grand mean. The formula for SS_{BG} is ...

$$SS_{BG}=\sum \left(\frac{x}{\overline{x}_G}\right)*(\overline{x}_G - \overline{x}_T)^2 = \sum n_G *(\overline{x}_G - \overline{x}_T)^2$$

... where SS_{BG} is the sum of squares between groups, x are the values, x-bar$_G$ are the group means, x-bar$_T$ is the grand mean, and n$_G$ are the sample size of the groups.

Sum of Squares Within Groups. The sum of the squared deviates within groups, SS_{WG}, is equal to the sum of the squared deviates of each value versus the group mean. The formula for SS_{WG} is ...

$$SS_{WG}=\sum (x - \overline{x}_G)^2$$

... where SS_{WG} is the sum of the squares within groups, x are the values, and x-bar$_G$ are the group means.

Degrees of Freedom Between Groups. The degrees of freedom between groups, DF_{BG}, is equal to the number of groups minus one.

Degrees of Freedom Within Groups. The degrees of freedom within groups, DF_{WG}, is equal to the total number of values minus the number of groups.

Mean Square Between Groups. The mean square between groups, MS_{BG}, is equal to the sum of squares between groups, SS_{BG}, divided by the degrees of freedom between groups, DF_{BG}.

Mean Square Within Groups. The mean square within groups, MS_{WG}, is equal to the sum of squares within groups, SS_{WG}, divided by the degrees of freedom within groups, DF_{WG}.

F Statistic. The f statistic is equal to the mean square between groups, MS_{BG}, — the explained variance— divided by the mean square within groups, MS_{WG} —the unexplained variance— with $DF=DF_{BG},DF_{WG}$.

For verifying results, the following formulas are also used in ANOVA.

Total Sum of Squares. The total sum of squares, SS_T, is equal to the sum of the squared deviates of each value versus the grand mean. The formula for SS_T is ...

$$SS_T = \sum (x - \bar{x}_T)^2$$

… where SS_T is the total sum of the squares, x are the values, and x-bar$_T$ is the grand mean.

FYI, $SS_T = SS_{BG} + SS_{WG}$.

Total Degrees of Freedom. The total degrees of freedom, DF_T, is equal to the total number of values minus one.

FYI, $DF_T = DF_{BG} + DF_{WG}$.

ANOVA Table

Oftentimes the results of ANOVA are shown as a table. For example, for the sample data shown in Table 40.1 on Page 171, Figure 40.1 shows the output of the Single Factor ANOVA test of the Data Analysis tool of Microsoft Excel 2010.

Figure 40.1

Anova: Single Factor

SUMMARY

Groups	Count	Sum	Average	Variance
Column 1	5	2634	526.8	10354.7
Column 2	5	3184	636.8	12304.2
Column 3	5	2085	417	6484

ANOVA

Source of Variation	SS	df	MS	F	P-value	F crit
Between Groups	120780.13	2	60390.067	6.2166154	0.0140347	3.8852938
Within Groups	116571.6	12	9714.3			
Total	237351.73	14				

Blank Page

One-Way ANOVA

Recall that:

- In a one-way ANOVA one compares a characteristic for one variable, called a factor, for three or more groups.
- In ANOVA one uses variances, which are the deviations of the values from the means.
- In ANOVA one uses the f statistic, which, in ANOVA, is the ratio of the explained variance to the unexplained variance.

Sample Data

Table 41.1 shows the SAT math scores for a random sample of five male students from three high schools where the factor is the school.

Table 41.1 (Male Students)		
School X	School Y	School Z
521	647	506
365	475	475
644	631	318
562	643	436
542	788	350

Null Hypothesis and Alternative Hypothesis

Say that one states that the average SAT scores of male students is equal for the three schools. That is, H_0: $\mu_X=\mu_Y=\mu_Z$ and the alternative hypothesis, H_1, is that there is a difference between the students' scores for at least two of the schools. Is it true?

Hypothesis Test

Although the one-way ANOVA is a bi-directional test, the sensitivity of the test allows one to view the test as a right-tail test. Recall that the critical value (CV) separates the acceptance region from the rejection region. Figure 41.1 on Page 178 shows that, if the test shows that the calculated f statistic is lower than the CV and, therefore, within the acceptance region, which is shown in gray, then, one fails to reject —that is, accepts— the null hypothesis. Conversely, if the test shows that the calculated f statistic is higher than the CV and, therefore, within the rejection region, which is shown in black, then, one rejects the null hypothesis.

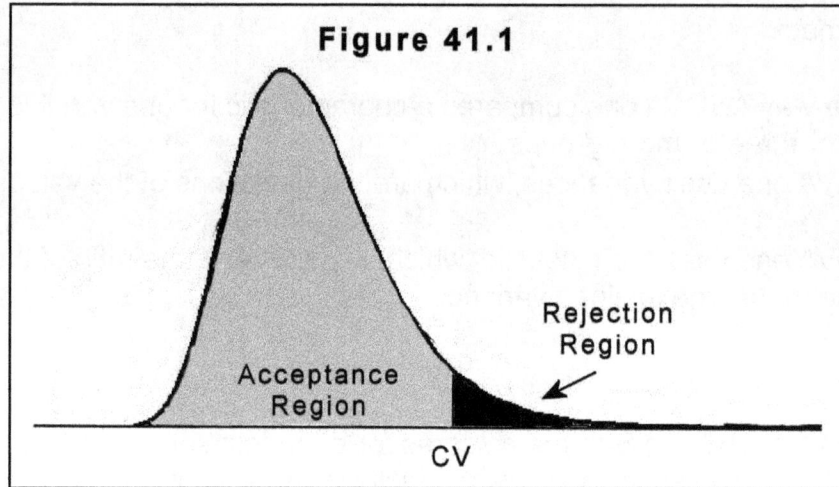

Figure 41.1

F Statistic

This section describes the steps to be followed for calculating the variances and the f statistic.

Group Means. Recall that the group means are equal to the sum of the values of the group divided by the sample size of the group. The group means for the sample data shown in Table 41.1 on Page 177 are 526.8, 636.8, and 417.0:

$$(521+365+...+542)/5=526.8$$
$$(647+475+...+788)/5=636.8$$
$$(506+475+...+350)/5=417.0$$

Grand Mean. Recall that the grand mean is equal to the sum of all the values divided by the total number of values. The grand mean for the sample data shown in Table 41.1 on Page 177 is 526.87:

$$(521+365+...+350)/15=526.87$$

Sum of Squares Between Groups. Recall that the sum of squared deviates between groups, SS_{BG}, is equal to the sum of the weighted differences between the group means and the grand mean. The SS_{BG} for the sample data shown in Table 41.1 on Page 177 is 120,780.1:

$$(5*(526.8-526.87)^2)+(5*(636.8-526.87)^2)+(5*(317.0-526.87)^2)=120,780.1$$

Sum of Squares Within Groups. Recall that the sum of squares within groups, SS_{WG}, is equal to the sum of the squared deviates of each value versus the group mean. The SS_{WG} for the sample data shown in Table 41.1 on Page 177 is 116,571.6:

$$(521-526.8)^2+(365-526.8)^2+...+(350-417.0)^2=116,571.6$$

Degrees of Freedom Between Groups. Recall that the degrees of freedom between groups, DF_{BG}, is equal to the number of groups minus one. The DF_{BG} for the sample data shown in Table 41.1 on Page 177 is 2:

$$3-1=2$$

Degrees of Freedom Within Groups. Recall that the degrees of freedom within groups, DF_{WG}, is equal to the total number of values minus the number of groups. The DF_{WG} for the sample data shown in Table 41.1 on Page 177 is 12:

$$15-3=12$$

Mean Square Between Groups. Recall that the mean square between groups, MS_{BG} —the explained variance— is equal to the sum of squares between groups, SS_{BG}, divided by the degrees of freedom between groups, DF_{BG}. The MS_{BG} for the sample data shown in Table 41.1 on Page 177 is 60,390.07:

$$120,780.1/2=60,390.07$$

Mean Square Within Groups. Recall that the mean square within groups, MS_{WG} —the unexplained variance— is equal to the sum of squares within groups (SS_{WG}) divided by the degrees of freedom within groups, DF_{WG}. The MS_{WG} for the sample data shown in Table 41.1 on Page 177 is 9,714.3:

$$116,571.6/12=9,714.3$$

F Statistic. Recall that the f statistic is equal to the mean square between groups, MS_{BG} —the explained variance— divided by the mean square within groups, MS_{WG} —the unexplained variance— with $DF=DF_{BG},DF_{WG}$. The f statistic for the sample data shown in Table 41.1 on Page 177 is 6.22 with DF=2,12:

$$60,390.07/9,714.3=6.22$$

Verification

This section describes the steps to be followed for verifying results.

Total Sum of Squares. Recall that the total sum of squares, SS_T, is equal to the sum of the squared deviates of each value versus the grand mean. The SS_T for the sample data shown in Table 41.1 on Page 177 is 237,351.73:

$$(521-526.87)^2+(365-526.87)^2+...+(350-526.87)^2=237,351.73$$

Also, $SS_T=SS_{BG}+SS_{WG}$:

$$120,780.1+116,571.6=237,351.7$$

Total Degrees of Freedom. Recall that the total degrees of freedom, DF_T, is equal to the total number of values minus one. The DF_T for the sample data shown in Table 41.1 on Page 177 is 14:

$$15-1=14$$

Also, $DF_T=DF_{BG}+DF_{WG}$:

$$2+12=14$$

ANOVA Table

Table 41.2 shows the ANOVA table for the sample data shown in Table 41.1 on Page 177.

Table 41.2				
Source	**SS**	**DF**	**MS**	**F**
Between	120,780.1	2	60,390.1	6.22
Within	116,571.6	12	9,714.3	
Total	237,351.7	14		

Critical Value and Confidence Level

Recall that if the test shows that the calculated f statistic is lower than the critical value (CV), then one fails to reject —that is, accepts— the null hypothesis. Conversely, if the test shows that the calculated f statistic is higher than the CV, then, one rejects the null hypothesis.

The f tables are used for finding the critical values (CVs) of f that correspond to a selected level of confidence, where CL=100%-α, which, for this example, 95% is selected. Table 41.3 shows, for $\alpha=5\%$ and for DF=2,12, that the f statistic that correspond to the 95% confidence level, which is CL=100%-5%, is 3.89. The confidence level shows the probability, which is 95%, that the f statistic is less than 3.89.

Table 41.3			
$\alpha=5\%$			
	$DF_1=1$	$DF_1=2$	$DF_1=\infty$
$DF_2=1$	161.45	199.50	254.31
$DF_2=12$	4.75	3.89	2.30
$DF_2=\infty$	3.84	3.00	1.00

Inference

The calculated f statistic of 6.22 is higher than the critical value (CV) of 3.89. therefore, one rejects the null hypothesis.

P Value

As an alternative to the traditional test, one may select an alpha level, say $\alpha=5\%$, and calculate the p=value. If $p<\alpha$ one rejects H_0. For example, using a web-based calculator, the right-tail probability of an f statistic of 6.22 with DF=2,12 is 0.01 (See Figure 41.2). [1] And, p=0.01 is less than 5%. Thus, one rejects H_0.

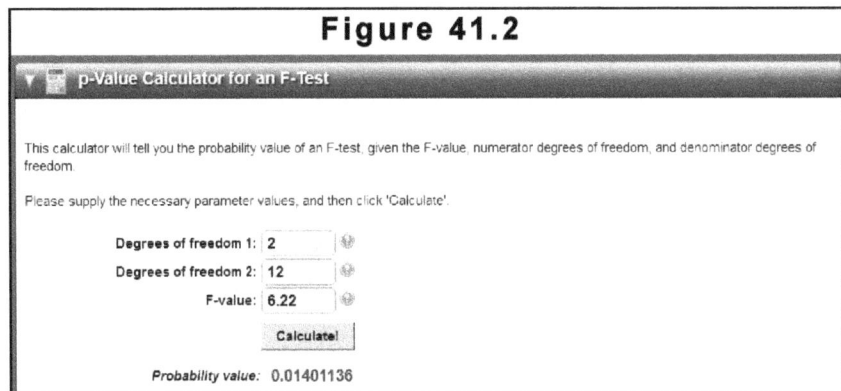

Figure 41.2

Microsoft Excel 2010

Figure 41.3 on Page 182 shows the output of the Single Factor ANOVA test of the Data Analysis tool of Microsoft Excel 2010.

Figure 41.3

Anova: Single Factor

SUMMARY

Groups	Count	Sum	Average	Variance
Column 1	5	2634	526.8	10354.7
Column 2	5	3184	636.8	12304.2
Column 3	5	2085	417	6484

ANOVA

Source of Variation	SS	df	MS	F	P-value	F crit
Between Groups	120780.13	2	60390.067	6.2166154	0.0140347	3.8852938
Within Groups	116571.6	12	9714.3			
Total	237351.73	14				

Other Tests

There are many post hoc tests, such as the Tukey test, that one can use to identify which samples differ.

Endnote

[1] Soper, D. (2015). *p-value calculator for an f-test*. URL:http://www.DanielSoper.com

Two-Way ANOVA Without Replication

Recall that in a two-way ANOVA without replication one compares a characteristic for two factors when there is only one value for each combination of factors. Chapter 41 described a one-factor ANOVA. This chapter describes a two-factor ANOVA without replication.

Sample Data

Table 42.1 shows the average scores of three math tests for five schools where the factors are school and test.

Table 42.1			
School	Test # 1	Test # 2	Test # 3
A	82	91	100
B	52	58	64
C	75	83	91
D	60	66	73
E	66	75	82

Null Hypothesis and Alternative Hypothesis

Say that one states that the school does not influence the scores, which is the null hypothesis, H_{0_1}. The alternative hypothesis, H_{1_1}, is that the school influences the scores. Is it true?

Say that one also states that the test does not influence the scores, which is the null hypothesis, H_{0_2}. The alternative hypothesis, H_{1_2}, is that the test influences the scores. Is it true?

Because of the lack of replicates, one cannot test the influence of the interaction between the test and the school.

Hypothesis Test

Although the one-way ANOVA is a bi-directional test, the sensitivity of the test allows one to view the test as a right-tail test. Recall that the critical value (CV) separates the acceptance region from the rejection region. Figure 42.1 on Page 184 shows that, if the test shows that the calculated f statistic is lower than the CV and, therefore, within the acceptance region, which is shown in gray, then, one fails to reject —that is, accepts— the null hypothesis. Conversely, if the test shows that the calculated f statistic is higher than the CV and, therefore, within the rejection region, which is shown in black, then, one rejects the null hypothesis.

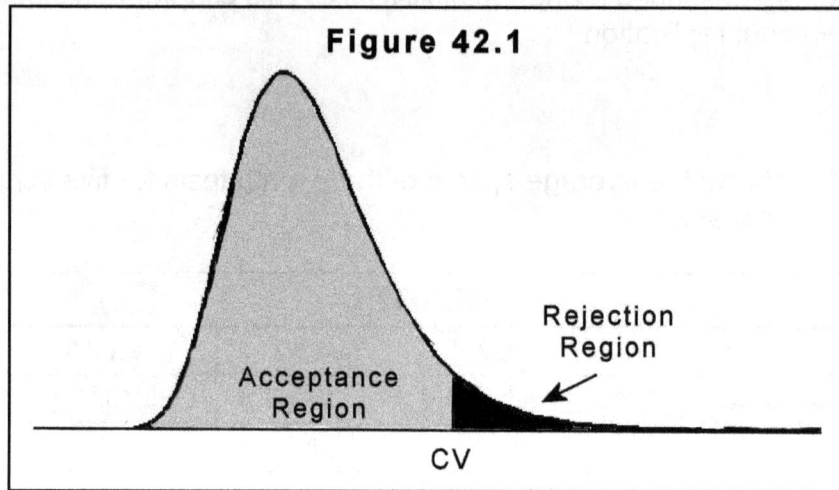

Figure 42.1

F Statistics

This section describes the steps to be followed for calculating the variances and the f statistics.

Table of Means. Calculate the means of each row, the means of each column, and the grand mean (See Table 42.2).

School (Rows)	Mean	Test (Columns)	Mean	Grand Mean
A	91.0	Test # 1	67.0	
B	58.0	Test # 2	74.6	
C	83.0	Test # 3	82	
D	66.3			
E	74.3			
				74.5

Table 42.2

Sum of Squares for the Rows. The formula for SS_{RWS} is …

$$SS_{RWS} = b * \sum(\bar{x}_{RWS} - \bar{x}_{TTL})^2$$

... where SS_{RWS} is the sum of squares for the rows, b is the number of columns, x_{RWS}-bar are the row means, and x_{TTL}-bar is the grand mean.

The sum of the squares for the rows, SS_{RWS}, is 2,050.4:

$$3*((91.0-74.5)^2+(58.0-74.5)^2+(83.0-74.5)^2+(66.4-74.5)^2+(74.3-74.5)^2)=2,050.4$$

Sum of Squares for the Columns. The formula for SS_{CLS} is ...

$$SS_{CLS} = a * \sum(\bar{x}_{CLS} - \bar{x}_{TTL})^2$$

... where SS_{CLS} is the sum of squares for the columns, a is the number of rows, x_{CLS}-bar are the column means, and x_{TTL}-bar is the grand mean.

The sum of the squares for the columns, SS_{CLS}, is 562.5:

$$5*((67.0-74.5)^2+(74.6-74.5)^2+(82.0-74.5)^2)=562.5$$

Sum of Squares for the Unexplained Variance. The formula for SS_{ERR} is ...

$$SS_{ERR} = \sum(x - \bar{x}_{RWS} - \bar{x}_{CLS} + \bar{x}_{TTL})^2$$

... where SS_{ERR} is the sum of squares for the error, x are the values, x_{RWS}-bar are the corresponding row means, x_{CLS}-bar are the corresponding column means, and x_{TTL}-bar is the grand mean.

The sum of the squares for the error, SS_{ERR}, is 12.8:

$$(82-91.0-67.0+74.5)^2+...+(82-74.3-82.0+74.5)^2=12.8$$

Degrees of Freedom for the Rows. The degrees of freedom for the rows, DF_{RWS}, is 4:

$$5-1=4$$

Degrees of Freedom for the Columns. The degrees of freedom for the columns, DF_{CLS}, is 2:

$$3\text{-}1=2$$

Degrees of Freedom for the Unexplained Variance. The degrees of freedom for the error, DF_{ERR}, is 8:

$$(5\text{-}1)*(3\text{-}1)=8$$

Mean Square for the Rows. The formula for MS_{RWS} is ...

$$MS_{RWS}=\frac{SS_{RWS}}{DF_{RWS}}$$

... where MS_{RWS} is the mean square for the rows, SS_{RWS} is the sum of the squares for the rows, and DF_{RWS} is the degrees of freedom for the rows.

The mean square for the rows, MS_{RWS}, is 512.6:

$$2,050.4/4=512.6$$

Mean Square for the Columns. The formula for MS_{CLS} is ...

$$MS_{CLS}=\frac{SS_{CLS}}{DF_{CLS}}$$

... where MS_{CLS} is the mean square for the columns, SS_{CLS} is the sum of the squares for the columns, and DF_{CLS} is the degrees of freedom for the columns.

The mean square for the columns, MS_{CLS}, is 281.3:

$$562.5/2=281.3$$

Mean Square for the Unexplained Variance. The formula for MS_{ERR} is ...

$$MS_{ERR}=\frac{SS_{ERR}}{DF_{ERR}}$$

... where MS_{ERR} is the mean square for the error, SS_{ERR} is the sum of the squares for the error, and DF_{ERR} is the degrees of freedom for the error.

The mean square for the error, MS_{ERR}, is 1.6:

$$12.8/8=1.6$$

F Statistic for the Rows. The formula for F_{RWS} is ...

$$F_{RWS}=\frac{MS_{RWS}}{MS_{ERR}}$$

... where F_{RWS} is the f statistic for the rows, MS_{RWS} is the mean square for the rows, and MS_{ERR} is the mean square for the error.

The f statistic for the rows, F_{RWS}, is 320.3:

$$512.6/1.6=320.3$$

F Statistic for the Columns. The formula for F_{CLS} is ...

$$F_{CLS}=\frac{MS_{CLS}}{MS_{ERR}}$$

... where F_{CLS} is the f statistic for the columns, MS_{CLS} is the mean square for the columns, and MS_{ERR} is the mean square for the error.

The f statistic for the columns, F_{CLS}, is X:

$$281.3/1.6=175.8$$

Verification

This section describes the steps to be followed for verifying results.

Total Sum of Squares. The formulas for SS_{TTL} are ...

$$SS_{TTL}=\sum(x-\bar{x}_{TTL})^2$$

... where SS_{TTL} is the total sum of squares, x are the values, and x_{TTL} is the grand mean.

And,

$$SS_{TTL} = SS_{RWS} + SS_{CLS} + SS_{ERR}$$

... where SS_{TTL} is the total sum of squares, SS_{RWS} is the sum of the squares for the rows, SS_{CLS} is the sum of the squares for the columns, and SS_{ERR} is the sum of the squares for the error.

The total sum of squares, SS_{TTL}, is X:

$$(82-74.5)^2+(52-74.5)^2+...+(82-74.5)^2=2,625.7$$

And,

$$2,050.4+562.5+12.8=2,625.7$$

Total Degrees of Freedom. The total degrees of freedom, DF_{TTL}, is 14:

$$15-1=14$$

And,

$$4+2+8=14$$

ANOVA Table

Table 42.3 shows the ANOVA table for the sample data shown in Table 42.1 on Page 183.

Table 42.3				
Source	**SS**	**DF**	**MS**	**F**
Rows	2,050.4	4	512.6	320.3
Columns	562.5	2	281.3	175.8
Error	12.8	8	1.6	
Total	2,625.7	14		

Critical Value and Confidence Level

Recall that if the test shows that the calculated f statistic is lower than the critical value (CV), then one fails to reject —that is, accepts— the null hypothesis. Conversely, if the test shows that the calculated f statistic is higher than the CV, then, one rejects the null hypothesis.

The f tables are used for finding the critical values (CVs) of f that correspond to a selected level of confidence, where CL=100%-α, which, for this example, 95% is selected. Table 42.4 shows, for α=5% and for DF=2,8 and DF=4,8 that the f statistics that correspond to the 95% confidence level, which is CL=100%-5%, are 4.46 and 3.84. The confidence level shows the probability, which is 95%, that the f statistics are less than 4.46 and 3.84.

Table 42.4		
α=5%		
	DF$_1$=2	DF$_1$=4
DF$_2$=8	4.46	
DF$_2$=8		3.84

Inferences

The calculated f statistics of 320.3 and 175.8 are both higher than the critical values (CV) of 3.84 and 4.46, therefore, one rejects both null hypotheses.

P Value

As an alternative to the traditional test, one may select an alpha level, say α=5%, and calculate the p=value. If p<α one rejects H$_0$.

For example, using a web-based calculator, the right-tail probability of an f statistic of 320.3 with DF=4,8 is nil (See Figure 42.2). [1] And, p is way less than 5%. Thus, one rejects H$_{0_1}$.

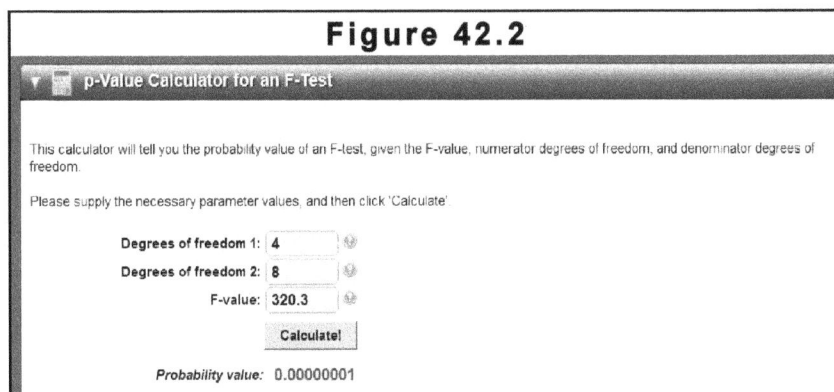

Figure 42.2

p-Value Calculator for an F-Test

This calculator will tell you the probability value of an F-test, given the F-value, numerator degrees of freedom, and denominator degrees of freedom.

Please supply the necessary parameter values, and then click 'Calculate'

Degrees of freedom 1: 4
Degrees of freedom 2: 8
F-value: 320.3
Calculate!

Probability value: 0.00000001

And, and the right-tail probability of an f statistic of 175.8 with DF=2,8 is nil (See Figure 42.3). [1] And, p is way less than 5%. Thus, one rejects H_{0_2}.

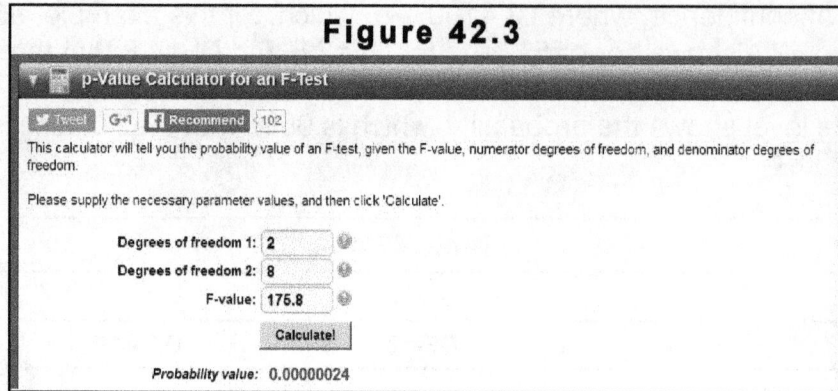

Figure 42.3

p-Value Calculator for an F-Test

Tweet   G+1   Recommend 102

This calculator will tell you the probability value of an F-test, given the F-value, numerator degrees of freedom, and denominator degrees of freedom.

Please supply the necessary parameter values, and then click 'Calculate'.

Degrees of freedom 1: 2
Degrees of freedom 2: 8
F-value: 175.8

Calculate!

Probability value: 0.00000024

Microsoft Excel 2010

Figure 42.4 shows the output of the Two Factor Without Replication ANOVA test of the Data Analysis tool of Microsoft Excel 2010.

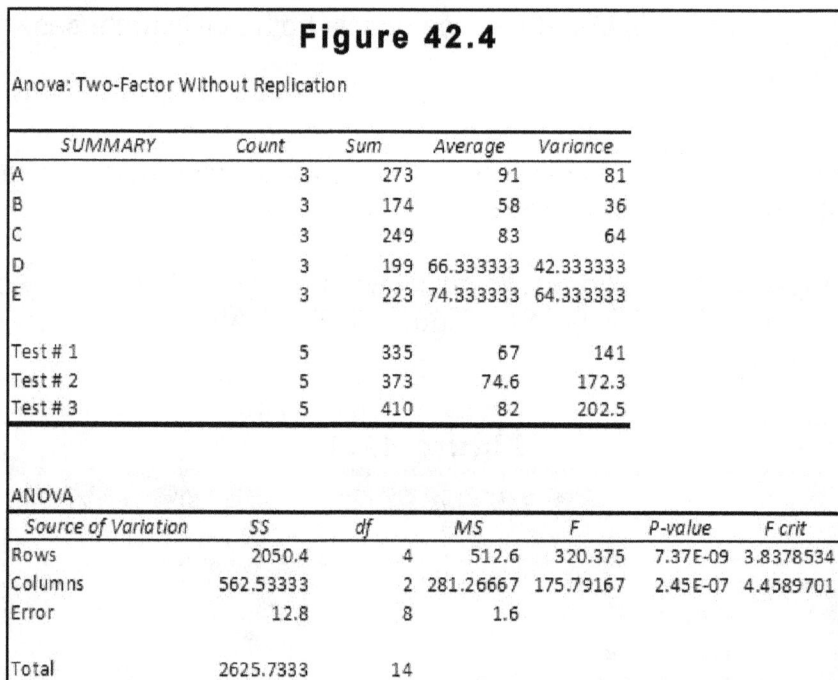

Figure 42.4

Anova: Two-Factor Without Replication

SUMMARY	Count	Sum	Average	Variance
A	3	273	91	81
B	3	174	58	36
C	3	249	83	64
D	3	199	66.333333	42.333333
E	3	223	74.333333	64.333333
Test # 1	5	335	67	141
Test # 2	5	373	74.6	172.3
Test # 3	5	410	82	202.5

ANOVA

Source of Variation	SS	df	MS	F	P-value	F crit
Rows	2050.4	4	512.6	320.375	7.37E-09	3.8378534
Columns	562.53333	2	281.26667	175.79167	2.45E-07	4.4589701
Error	12.8	8	1.6			
Total	2625.7333	14				

Endnote

[1] Soper, D. (2015). *p-value calculator for an f-test*. URL:http://www.DanielSoper.com

Two-Way ANOVA With Replication

Recall that in a two-way ANOVA with replication one compares a characteristic for two factors when there is more than one value for each combination of factors. Chapter 42 described a two-factor ANOVA without replication. This chapter describes a two-factor ANOVA with replication.

Sample Data

Table 43.1 shows shows the IQs of five males and five females from three different countries where the factors are gender and country-of-origin.

Table 43.1			
	HK	**USA**	**Mex**
Male	79	106	95
	111	91	100
	87	97	57
	107	104	100
	159	99	79
Female	115	84	98
	115	98	78
	114	100	59
	98	83	78
	92	94	84

Null Hypothesis and Alternative Hypothesis

Say that one states that the gender does not influence the IQs which is the null hypothesis, H_{0_1}. The alternative hypothesis, H_{1_1}, is that the gender influences the scores. Is it true?

Next, say that one also states that the country-of-origin does not influence the IQs, which is the null hypothesis, H_{0_2}. The alternative hypothesis, H_{1_2}, is that the country-of-origin influences the scores. Is it true?

Finally, say that one also states that the interaction between gender and country-of-origin does not influence the IQs, which is the null hypothesis, H_{0_3}. The alternative hypothesis, H_{1_3}, is that the country-of-origin influences the scores. Is it true?

Hypothesis Test

Although the one-way ANOVA is a bi-directional test, the sensitivity of the test allows one to view the test as a right-tail test. Recall that the critical value (CV) separates the acceptance region from the rejection region. Figure 43.1 shows that, if the test shows that the calculated f statistic is lower than the CV and, therefore, within the acceptance region, which is shown in gray, then, one fails to reject —that is, accepts — the null hypothesis. Conversely, if the test shows that the calculated f statistic is higher than the CV and, therefore, within the rejection region, which is shown in black, then, one rejects the null hypothesis.

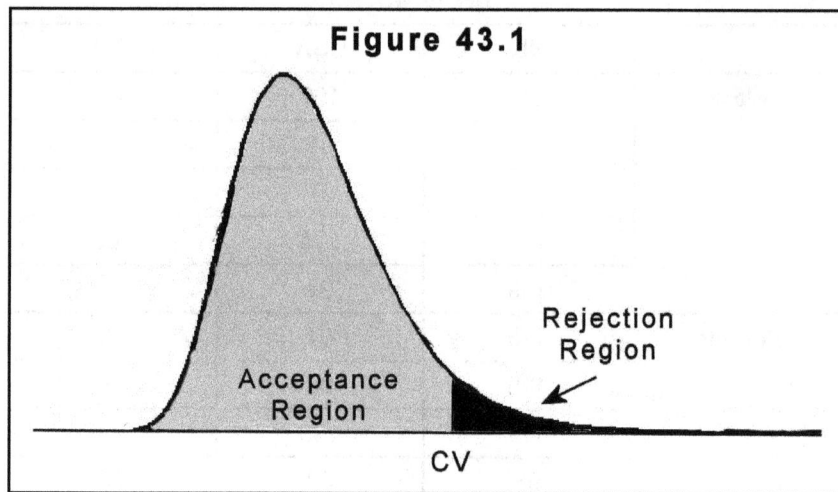

Figure 43.1

F Statistics

This section describes the steps to be followed for calculating the variances and the f statistics.

Table of Means. Calculate the means of each group, the means of each row, the means of each column, and the grand mean (See Table 43.2).

Table 43.2			
	HK	**USA**	**MEX**	**Total**
Male	108.6	99.4	86.2	98.1
Female	106.8	91.8	79.4	92.7
Total	107.7	95.6	82.8	95.4

Sum of Squares for the Rows. The formula for SS_{RWS} is …

$$SS_{RWS} = r * b * \sum (\overline{x}_{RWS} - \overline{x}_{TTL})^2$$

… where SS_{RWS} is the sum of squares for the rows, r is the number of replicates, b is the number of columns, x_{RWS}-bar are the row means, and x_{TTL}-bar is the grand mean.

The sum of the squares for the rows, SS_{RWS}, is 218.7:

$$5*3*((98.1-95.4)^2+(92.7-95.4)^2)=218.7$$

Sum of Squares for the Columns. The formula for SS_{CLS} is …

$$SS_{CLS} = r * a * \sum (\overline{x}_{CLS} - \overline{x}_{TTL})^2$$

… where SS_{CLS} is the sum of squares for the columns, r is the number of replicates, a is the number of rows, x_{CLS}-bar are the column means, and x_{TTL}-bar is the grand mean.

The sum of the squares for the columns, SS_{CLS}, is 3,100.9:

$$5*2*((107.7-95.4)^2+(95.6-95.4)^2+(82.8-95.4)^{2+})=3,100.9$$

Sum of Squares for RxC. The formula for SS_{RxC} is …

$$SS_{RxC} = r * \sum (\overline{x}_{GRP} - \overline{x}_{RWS} - \overline{x}_{CLS} + \overline{x}_{TTL})^2$$

… where SS_{RxC} is the sum of squares for the interaction between rows and columns, r is the number of replicates, x_{GRP}-bar are the group means, x_{RWS}-bar are the corresponding row means,,x_{CLS}-bar are the corresponding column means, and x_{TTL}-bar is the grand mean.

The sum of the squares for the interaction between rows and columns, SS_{RxC}, is 49.4:

$$5*((108.6-92.7-106.8+95.4)^2+...+(79.4-92.7-79.4+95.4)^2)=49.4$$

Sum of Squares for the Unexplained Variance. The formula for SS_{ERR} is ...

$$SS_{ERR}=\sum(x-\bar{x}_{RWS}-\bar{x}_{CLS}+\bar{x}_{TTL})^2$$

... where SS_{ERR} is the sum of squares for the error, x are the values, x_{RWS}-bar are the corresponding row means, x_{CLS}-bar are the corresponding column means, and x_{TTL}-bar is the grand mean.

The sum of the squares for the error, SS_{ERR}, is 6,914:

$$(79-108.6)^2+...+(84-79.4)^2=6,914$$

Degrees of Freedom for the Rows. The degrees of freedom for the rows, DF_{RWS}, is 1:

$$2-1=1$$

Degrees of Freedom for the Columns. The degrees of freedom for the columns, DF_{CLS}, is 2:

$$3-1=2$$

Degrees of Freedom for RxC. The degrees of freedom for the columns, DF_{RxC}, is 2:

$$(2-1)*(3-1)=2$$

Degrees of Freedom for the Unexplained Variance. The degrees of freedom for the error, DF_{ERR}, is 24:

$$(5-1)*(2)*(3)=24$$

Mean Square for the Rows. The formula for MS_{RWS} is ...

$$MS_{RWS}=\frac{SS_{RWS}}{DF_{RWS}}$$

... where MS_{RWS} is the mean square for the rows, SS_{RWS} is the sum of the squares for the rows, and DF_{RWS} is the degrees of freedom for the rows.

The mean square for the rows, MS_{RWS}, is 218.7:

$$218.7/1 = 218.7$$

Mean Square for the Columns. The formula for MS_{CLS} is ...

$$MS_{CLS} = \frac{SS_{CLS}}{DF_{CLS}}$$

... where MS_{CLS} is the mean square for the columns, SS_{CLS} is the sum of the squares for the columns, and DF_{CLS} is the degrees of freedom for the columns.

The mean square for the columns, MS_{CLS}, is 1,550.5:

$$3,100.9/2 = 1,550.5$$

Mean Square for RxC. The formula for MS_{RxC} is ...

$$MS_{RxC} = \frac{SS_{RxC}}{DF_{RxC}}$$

... where MS_{RxC} is the mean square for RxC, SS_{RxC} is the sum of the squares for RxC, DF_{RxC} is the degrees of freedom for RxC.

The mean square for RxC, MS_{RxC}, is 24.7:

$$49.4/2 = 24.7$$

Mean Square for the Unexplained Variance. The formula for MS_{ERR} is ...

$$MS_{ERR} = \frac{SS_{ERR}}{DF_{ERR}}$$

... where MS_{ERR} is the mean square for the error, SS_{ERR} is the sum of the squares for the error, and DF_{ERR} is the degrees of freedom for the error.

The mean square for the error, MS_{ERR}, is 288.1:

$$6,914/24 = 288.1$$

F Statistic for the Rows. The formula for F_{RWS} is ...

$$F_{RWS} = \frac{MS_{RWS}}{MS_{ERR}}$$

... where F_{RWS} is the f statistic for the rows, MS_{RWS} is the mean square for the rows, and MS_{ERR} is the mean square for the error.

The f statistic for the rows, F_{RWS}, is 0.76:

$$218.7/288.1=0.76$$

F Statistic for the Columns. The formula for F_{CLS} is ...

$$F_{CLS} = \frac{MS_{CLS}}{MS_{ERR}}$$

... where F_{CLS} is the f statistic for the columns, MS_{CLS} is the mean square for the columns, and MS_{ERR} is the mean square for the error.

The f statistic for the columns, F_{CLS}, is 5.39:

$$1,550.5/288.1=5.39$$

F Statistic for RxC. The formula for F_{RxC} is ...

$$F_{RxC} = \frac{MS_{RxC}}{MS_{ERR}}$$

... where F_{RxC} is the f statistic for RxC, MS_{RxC} is the mean square for RxC, and MS_{ERR} is the mean square for the error.

The f statistic for the columns, F_{RxC}, is 0.09:

$$24.7/288.1=0.09$$

Verification

This section describes the steps to be followed for verifying results.

Total Sum of Squares. The formulas for SS_{TTL} are ...

$$SS_{TTL} = \sum (x - \bar{x}_{TTL})^2$$

... where SS_{TTL} is the total sum of squares, x are the values, and x_{TTL} is the grand mean.

And,

$$SS_{TTL} = SS_{RWS} + SS_{CLS} + SS_{RxC} + SS_{ERR}$$

... where SS_{TTL} is the total sum of squares, SS_{RWS} is the sum of the squares for the rows, SS_{CLS} is the sum of the squares for the columns, SS_{RxC} is the sum of the squares for RxC, and SS_{ERR} is the sum of the squares for the error.

The total sum of squares, SS_{TTL}, is 10,283:

$$(79-95.4)^2+(111-95.4)^2+...+(84-95.4)^2=10,283$$

And,

$$218.7+3,100.9+49.4+6,914=10,283$$

Total Degrees of Freedom. The total degrees of freedom, DF_{TTL}, is 29:

$$30-1=29$$

And,

$$1+2+2+24=29$$

ANOVA Table

Table 43.3 on Page 198 shows the ANOVA table for the sample data shown in Table 43.1 on Page 191.

Table 43.3				
Source	SS	DF	MS	F
Rows	218.7	1	218.7	0.76
Columns	3,100.9	2	1,550.5	5.39
RxC	49.4	2	24.7	0.09
Error	6,914.0	24	288.1	
Total	10,283.0	29		

Critical Value and Confidence Level

Recall that if the test shows that the calculated f statistic is lower than the critical value (CV), then one fails to reject —that is, accepts— the null hypothesis. Conversely, if the test shows that the calculated f statistic is higher than the CV, then, one rejects the null hypothesis.

The f tables are used for finding the critical values (CVs) of f that correspond to a selected level of confidence, where CL=100%-α, which, for this example, 95% is selected. Table 42.4 shows, for α=5% and for DF=1,24 and DF=2,24, that the f statistics that correspond to the 95% confidence level, which is CL=100%-5%, are 4.26 and 3.40. The confidence level shows the probability, which is 95%, that the f statistics are less than 4.26 and 3.4.

Table 42.4		
α=5%		
	$DF_1=1$	$DF_1=2$
$DF_2=24$	4.26	3.40

Inferences

First, the calculated f statistic of 0.76 is lower than the critical values (CV) of 4.26, therefore, one fails to reject —that is, accepts— H_{0_1}.

Second, the calculated f statistic of 5.39 is higher than the critical values (CV) of 3.40, therefore, one rejects H_{0_2}.

Third, the calculated f statistic of 0.09 is lower than the critical values (CV) of

3.40, therefore, one fails to reject —that is, accepts— H_{0_1}.

P Value

As an alternative to the traditional test, one may select an alpha level, say $\alpha=5\%$, and calculate the p=value. If $p<\alpha$ one rejects H_0.

For example, using a web-based calculator, the right-tail probability of an f statistic of 0.76 with DF=1,24 is 39.2% (See Figure 43.2). [1] And, p is higher than 5%. Thus, one fails to reject —that is, accepts— H_{0_1}.

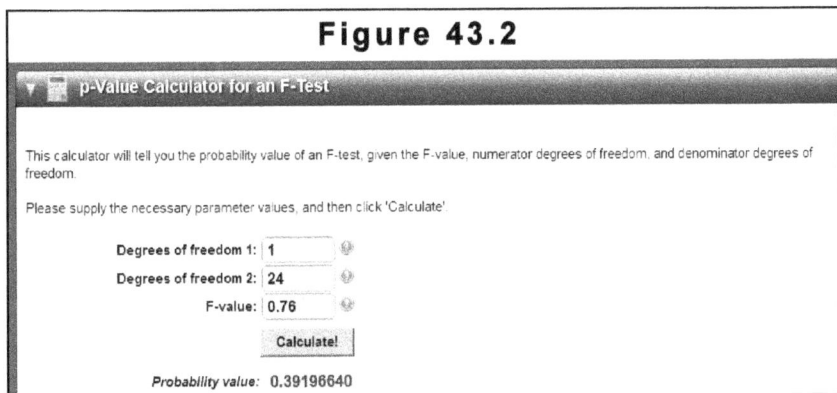

Figure 43.2

p-Value Calculator for an F-Test

This calculator will tell you the probability value of an F-test, given the F-value, numerator degrees of freedom, and denominator degrees of freedom.

Please supply the necessary parameter values, and then click 'Calculate'

Degrees of freedom 1: 1
Degrees of freedom 2: 24
F-value: 0.76

Calculate!

Probability value: 0.39196640

And, the right-tail probability of an f statistic of 5.39 with DF=2,24 is 1.2% (See Figure 43.3). [1] And, p is less than 5%. Thus, one rejects H_{0_2}.

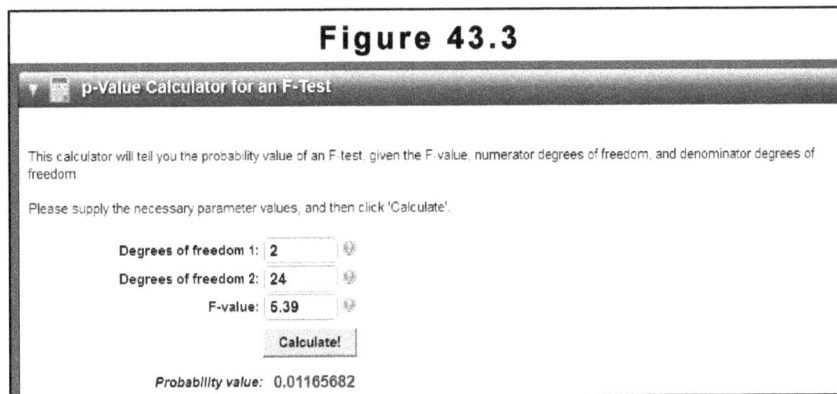

Figure 43.3

p-Value Calculator for an F-Test

This calculator will tell you the probability value of an F-test, given the F-value, numerator degrees of freedom, and denominator degrees of freedom.

Please supply the necessary parameter values, and then click 'Calculate'.

Degrees of freedom 1: 2
Degrees of freedom 2: 24
F-value: 5.39

Calculate!

Probability value: 0.01165682

Finally, the right-tail probability of an f statistic of 0.09 with DF=2,24 is 91.4% (See Figure 43.4 on Page 200). [1] And, p is higher than 5%. Thus, one fails to reject — that is, accepts— H_{0_3}.

Figure 43.4

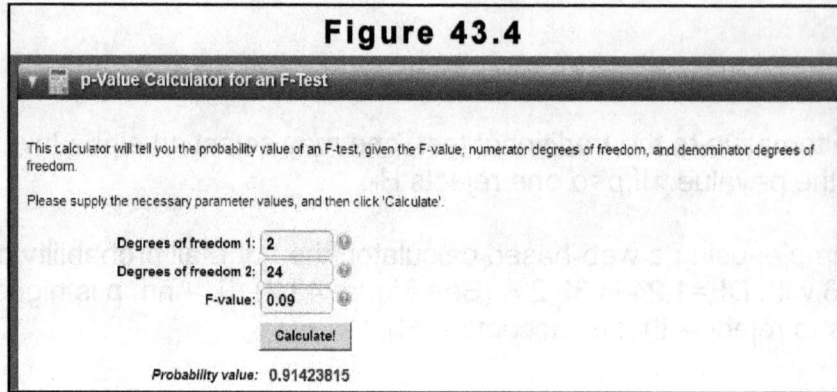

p-Value Calculator for an F-Test

This calculator will tell you the probability value of an F-test, given the F-value, numerator degrees of freedom, and denominator degrees of freedom.

Please supply the necessary parameter values, and then click 'Calculate'.

Degrees of freedom 1: 2
Degrees of freedom 2: 24
F-value: 0.09

Calculate!

Probability value: 0.91423815

Microsoft Excel 2010

Figure 43.5 on Page 201 shows the output of the Two Factor Without Replication ANOVA test of the Data Analysis tool of Microsoft Excel 2010.

Figure 43.5

Anova: Two-Factor With Replication

SUMMARY	HK	USA	MEX	Total
Male				
Count	5	5	5	15
Sum	543	497	431	1471
Average	108.6	99.4	86.2	98.066667
Variance	972.8	35.3	340.7	475.92381
Female				
Count	5	5	5	15
Sum	534	459	397	1390
Average	106.8	91.8	79.4	92.666667
Variance	120.7	62.2	196.8	242.95238
Total				
Count	10	10	10	
Sum	1077	956	828	
Average	107.7	95.6	82.8	
Variance	486.9	59.377778	251.73333	

ANOVA

Source of Variation	SS	df	MS	F	P-value	F crit
Sample	218.7	1	218.7	0.7591553	0.3922256	4.2596773
Columns	3100.8667	2	1550.4333	5.3818918	0.0117222	3.4028261
Interaction	49.4	2	24.7	0.0857391	0.9181135	3.4028261
Within	6914	24	288.08333			
Total	10282.967	29				

Endnote

[1] Soper, D. (2015). *p-value calculator for an f-test*. URL:http://www.DanielSoper.com

Blank Page

Wilcoxon Rank-Sum Test

Chapter 32 described a t test for two independent means. This chapter describes its non-parametric equivalent: the Wilcoxon rank-sum test.

Sample Data

In the United States, credit card debt is not normally distributed, it is right-skewed. Table 44.1 shows shows the credit card debt of a random sample of ten males and ten females.

Table 44.1			
Male		**Female**	
0	0	0	0
0	0	250	250
0	2,500	0	2,200
3,400	0	0	350
0	0	1,000	100

Null Hypothesis and Alternative Hypothesis

Say that one states that the median debt for males is equal to the median debt for females. That is, H_0: $M_M = M_F$ and H_1: $M_M \neq M_F$. Is it true?

Hypothesis Test

Recall that the critical value (CV) separates the acceptance region from the rejection region. Figure 44.1 on Page 204 shows that, if the test shows that the calculated z score is lower than the CV and, therefore, within the acceptance region, which is shown in gray, then, one fails to reject —that is, accepts— the null hypothesis. Conversely, if the test shows that the calculated z score is higher than the CV and, therefore, within the rejection region, which is shown in black, then, one rejects the null hypothesis.

Figure 44.1

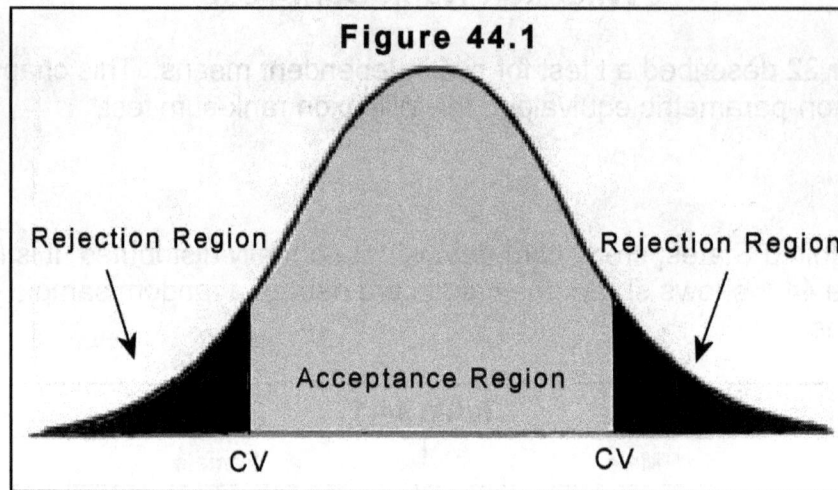

Ranks

Table 44.2 shows the debt ranked from low to high.

Table 44.2					
Male			**Female**		
Debt	**Start Rank**	**End Rank**	**Debt**	**Start Rank**	**End Rank**
0	1	6.5*	0	9	6.5*
0	2	6.5*	0	10	6.5*
0	3	6.5*	0	11	6.5*
0	4	6.5*	0	12	6.5*
0	5	6.5*	100	13	13
0	6	6.5*	250	14	14.5 **
0	7	6.5*	250	15	14.5 **
0	8	6.5*	350	16	16
2,500	19	19	1,000	17	17
3,400	20	20	2,200	18	18
Total (W₁)		91	**Total (W₂)**		119

* Rank=(1+2+...+12)/12=6.5

** Rank=(14+15)/2=14.5

Z Score

This section describes the steps to be followed for calculating the z score.

Mean. The formula for the mean, μ_W, is …

$$\mu_W = \frac{n_1 * (n_1 + n_1 + 1)}{2}$$

… where μ_W is the mean, n_1 is the sample size for the first sample, and n_2 is the sample size for the second sample.

For the first sample, the relevant mean is 105:

$$10*(10+10+1)/2=105$$

Standard Deviation. The formula for the standard deviation, σ_W, is …

$$\sigma_W = \sqrt{\frac{n_1 * n_2 * (n_1 + n_1 + 1)}{12}}$$

… where σ_W, is the standard deviation, n_1 is the sample size for the first sample, and n_2 is the sample size for the second sample.

For the first sample, the relevant standard deviation is 13,2:

$$SQRT((10*10*(10+10+1))/12)=13.2$$

Z Score. The formula for the z score is …

$$z = \frac{W - \mu_W}{\sigma_W}$$

… where z is the z score, μ_W is the mean, and σ_W, is the standard deviation.

For the first sample, the relevant z score is -1.06:

$$(91-105)/13.2=-1.06$$

Verification

This section describes the steps to be followed for verifying results.

Mean. For the second sample, the relevant mean is 105:

$$10*(10+10+1)/2=105$$

Standard Deviation. For the second sample, the relevant standard deviation is 13,2:

$$SQRT((10*10*(10+10+1))/12)=13.2$$

Z Score. For the second sample, the relevant z score is 1.06:

$$(119-105)/13.2=1.06$$

Critical Value and Confidence Level

Recall that if the test shows that the calculated z score is lower than the critical value (CV), then one fails to reject —that is, accepts— the null hypothesis. Conversely, if the test shows that the calculated z score is higher than the CV, then, one rejects the null hypothesis.

The z table is used for finding the critical values (CVs) of z that correspond to a selected level of confidence, where CL=100%-α, which, for this example, 95% is selected. Table 44.3 and Figure 44.2 on Page 207 show that the z scores that correspond to the 95% confidence level, which is CL=100%-5%, are -1.96 for the left side and +1.96 for the right side. The confidence level shows the probability, which is 95%, that the z scores are between -1.96 and +1.96. Note that $\alpha/2$=2.5%.

Table 44.3		
z Score		Cumulative Left-Tail Probability
-1.9	.60	0.02750
+1.9	.60	0.97500

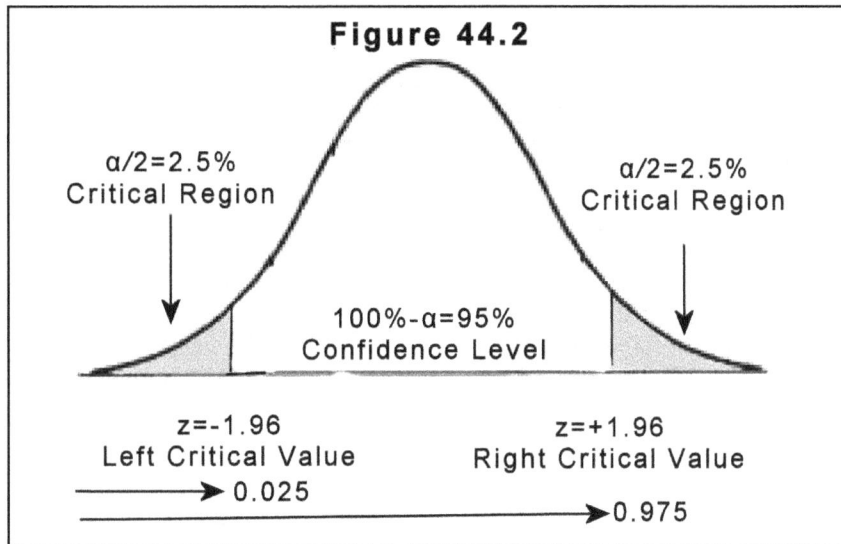

Figure 44.2

$\alpha/2 = 2.5\%$ Critical Region

$\alpha/2 = 2.5\%$ Critical Region

$100\% - \alpha = 95\%$ Confidence Level

$z = -1.96$ Left Critical Value → 0.025

$z = +1.96$ Right Critical Value → 0.975

Inference

The calculated z score of -1.06 is lower than the critical values (CV) of -1.96, therefore, one fails to reject —that is, accepts— H_0.

Statdisk

Figure 44.3 and Figure 44.4 on Page 208 show the output of Statdisk. [2]

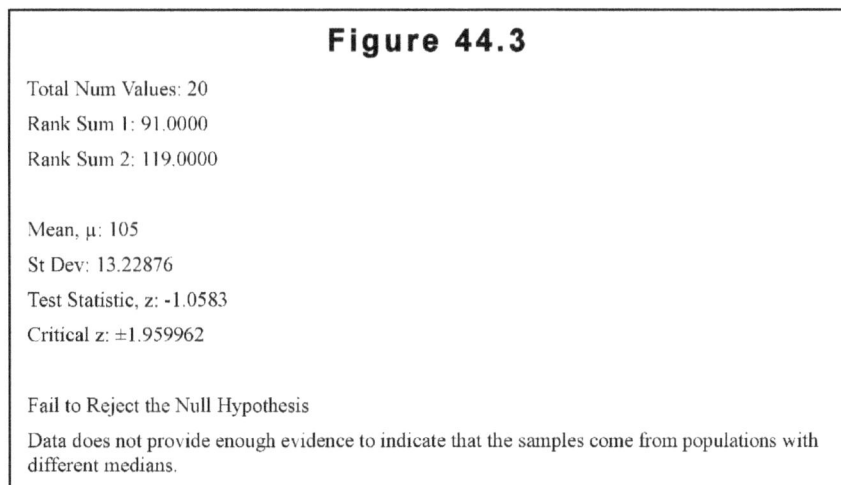

Figure 44.3

Total Num Values: 20

Rank Sum 1: 91.0000

Rank Sum 2: 119.0000

Mean, μ: 105

St Dev: 13.22876

Test Statistic, z: -1.0583

Critical z: ±1.959962

Fail to Reject the Null Hypothesis

Data does not provide enough evidence to indicate that the samples come from populations with different medians.

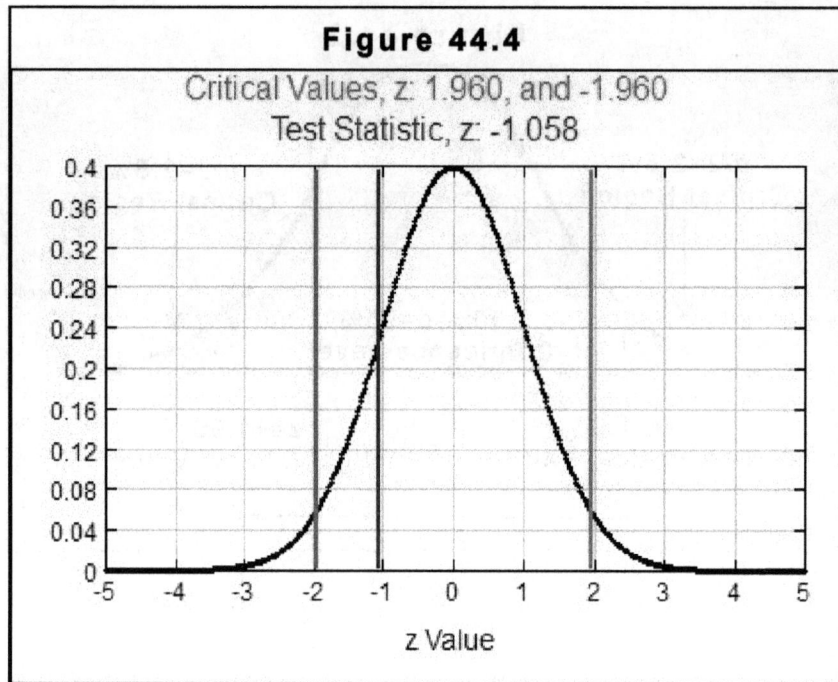

Figure 44.4

Critical Values, z: 1.960, and -1.960
Test Statistic, z: -1.058

Endnotes

[1] Sorensen, M. (2013). Gender vs. credit card debt. Hoboken, NJ: Statcrunch / Pearson Education. URL: http://www.statcrunch.com

[2] Triola, M. (2009). Statdisk [Computer software]. Hoboken, NJ: Pearson Education.

Wilcoxon Signed-Rank Test

Chapter 33 described a t test for two paired means. This chapter describes its non-parametric equivalent: the Wilcoxon signed-rank test.

Sample Data

In the United States, income is not normally distributed, it is right-skewed. Table 45.1 shows the hourly wages of ten factory workers ten years ago and now as well as the differences.

Table 45.1									
2006	8.0	8.0	7.9	7.8	8.1	8.0	7.9	7.8	8.10	7.9
2016	10.0	7.0	10.1	6.8	9.9	9.8	6.9	10.0	7.10	9.9
Dif.	+2	-1	+2.2	-1	+1.8	+1.8	-1	+2.2	-1	+2

Null Hypothesis and Alternative Hypothesis

Say that one states that the median wages now are higher than the median wages ten years ago. That is, H_0: $M_A-M_B=0$ and H_1: $M_A-M_B>0$. Is it true?

Hypothesis Test

Wilkinson tables are used for finding the critical value (CV) of the T statistic that correspond to a selected level of confidence (CL), where CL=100%-α, which is the significance level. If the T statistics is greater than the CV, one fails to reject —that is, accepts— the null hypothesis. Conversely, if the T statistics is equal to or less than the CV, one rejects the null hypothesis.

Ranks

Table 45.3 on Page 210 shows the differences ranked from low to high.

Table 45.2					
Dif.	Start Rank	End Rank	Sign	+W	-W
-1	1	2.5 [a]	-		2.5
-1	2	2.5 [a]	-		2.5
-1	3	2.5 [a]	-		2.5
-1	4	2.5 [a]	-		2.5
+1.8	5	5.5 [b]	+	5.5	
+1.8	6	5.5 [b]	+	5.5	
+2	7	7.5 [c]	+	7.5	
+2	8	7.5 [c]	+	7.5	
+2.2	9	9.5 [d]	+	9.5	
+2.2	10	9.5 [d]	+	9.5	
Total				45	10

a Rank=(1+2+3+4)/4=2.5
b Rank=(5+6)/2=5.5
c Rank=(7+8)/2=7.5
d Rank=(9+10)/2=9.5

Sample Size

The sample size is equal to the number of non-zero differences. Therefore the sample size is 10.

Test Statistic

This section describes the steps to be followed for calculating the test statistic.

W+ and W-. Table 45.2 shows the totals for W+ and W-, which are 45 and 10, respectively.

Wilkinson's T Statistic. The Wilkinson T statistic is the smallest of W+ and W-. Therefore, the Wilkinson T statistic is10.

Verification

The formula for verifying the results, W_{TOTAL}, is ...

$$W_{TOTAL} = W_{PLUS} + W_{MINUS} = \frac{n*(n+1)}{2}$$

... where W_{TOTAL}, W_{PLUS}, and W_{MINUS} are the rank sums and n is sample size.

The total rank sum, W_{TOTAL}, is 55:

$$45 + 10 = (10*(10+1))/2 = 55$$

Critical Value and Confidence Level

Recall that if the T statistics is greater than the CV, one fails to reject —that is, accepts— the null hypothesis. Conversely, if the T statistics is equal to or less than the CV, one rejects the null hypothesis.

Also recall that Wilkinson tables are used for finding the critical value (CV) of the T statistic that correspond to a selected level of confidence (CL), where CL=100%-α, which is the significance level. For the example, α=5% is selected and, therefore, CL=100%-5% or 95%.

Table 45.3 show that the Wilkinson T statistic that correspond to the 95% confidence level, which is CL=100%-5%, is 10.

Table 45.3		
N	5.00%	
	One-Tail	Two-Tail
10	10	8

Inference

The calculated Wilkinson T score of 10 is equal to or lower than the critical value (CV) of 10, therefore, one rejects H_0.

Social Science Statistics

Figure 45.1 on Page 212 show the output of the Social Science Statistics'

Wilcoxon Signed-Rank Test Calculator. [1]

Figure 45.1

Wilcoxon Signed-Rank Test Calculator

Note: You can find further information about this calculator, here.

Success!

Explanation of results

We have calculated both a W-value and Z-value. If the size of N is at least 20 - see the Results Details box - then the distribution of the Wilcoxon W statistic tends to form a normal distribution. This means you can use the Z-value to evaluate your hypothesis. If, on the other hand, the size of N is low, and particularly if it's below 10, you should use the W-value to evaluate your hypothesis.

You should also note that if a subject's difference score is zero - that is, if a subject has the same score in both treatment conditions - then the test discards the individual from the analysis and reduces the sample size. If you have a lot of ties, this procedure will undermine the reliability of the test (and also suggests that the requirement that the data is continuous has not been met).

Treatment 1	Treatment 2	Sign	Abs	R	Sign R
8	10	-1	2	7.5	-7.5
8	7	1	1	2.5	2.5
7.9	10.1	-1	2.2	9	-9
7.8	6.8	1	1	2.5	2.5
8.1	9.9	-1	1.8	5.5	-5.5
8	9.8	-1	1.8	5.5	-5.5
7.9	6.9	1	1	2.5	2.5
7.8	10	-1	2.2	10	-10
8.1	7.1	1	1	2.5	2.5
7.9	9.9	-1	2	7.5	-7.5

Significance Level:

○ 0.01

◉ 0.05

1 or 2-tailed hypothesis?:

◉ One-tailed

○ Two-tailed

Result Details

W-value: 10
Mean Difference: 0.95
Sum of pos. ranks: 10
Sum of neg. ranks: 45

Z-value: -1.7838
Mean (W): 27.5
Standard Deviation (W): 9.81

Sample Size (N): 10

Result 1 - Z-value

The Z-value is -1.7838. The p-value is 0.03754. The result is significant at $p \leq 0.05$.

Result 2 - W-value

The W-value is 10. The critical value of W for $N = 10$ at $p \leq 0.05$ is 10. Therefore, the result is significant at $p \leq 0.05$.

Calculate	Reset

Endnote

[1] Social Science Statistics. (n.d.). *Wilcoxon Signed-Rank Test Calculator*. URL: http://www.socscistatistics.com

Kruskal-Wallis Test

Chapter 41 described a one-way ANOVA test. This chapter describes its non-parametric equivalent: the Kruskal-Wallis test.

Sample Data

In the United States, income is not normally distributed, it is right-skewed. Table 46.1 shows shows the median annual household income of a random sample of five households in each of three states.

Table 46.1		
California	Florida	Texas
60.5	46.1	53.9
66.6	50.7	59.3
63.5	48.4	56.6
57.5	43.8	51,2
54.5	41.5	48.5

Null Hypothesis and Alternative Hypothesis

Say that one states that the median income is equal for all three states. That is, H_0: $M_{CA}=M_{FL}=M_{TX}$ and the alternative hypothesis, H_1, is that there is a difference between the median income for at least two of the states. Is it true?

Hypothesis Test

Recall that the critical value (CV) separates the acceptance region from the rejection region. Figure 46.1 on Page 214 shows that, if the test shows that the calculated test statistic, is lower than the CV and, therefore, within the acceptance region, which is shown in gray, then, one fails to reject —that is, accepts— the null hypothesis. Conversely, if the test shows that the calculated test statistic is higher than the CV and, therefore, within the rejection region, which is shown in black, then, one rejects the null hypothesis.

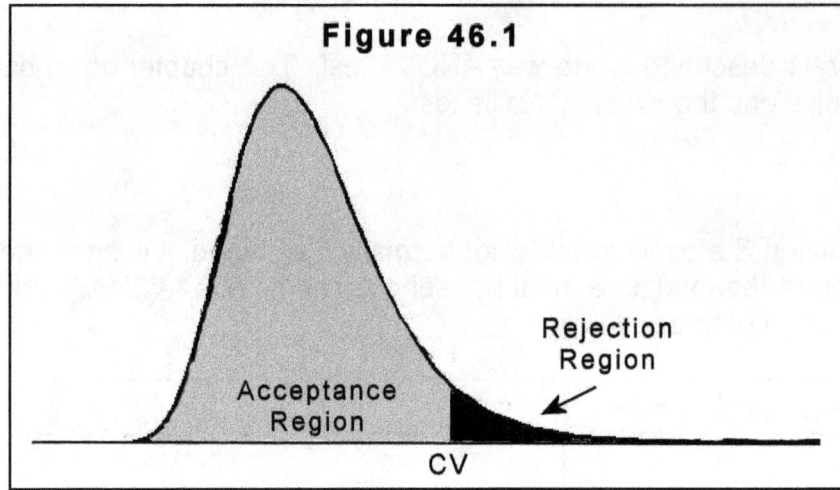

Figure 46.1

Ranks

Table 44.2 shows the debt ranked from low to high.

Table 46.2

California		Florida		Texas	
Inc.	**Rank**	**Inc.**	**Rank**	**Inc.**	**Rank**
54.5	9	41.5	1	48.5	5
57.5	11	43.8	2	51.2	7
60.5	13	46.1	3	53.9	8
63.5	14	48.4	4	56.6	10
66.6	15	50.7	6	59.3	12
Total (R_1)	62	**Total (R_2)**	16	**Total (R_3)**	42

Note. As shown in Chapter 44 and Chapter 45, the rank of tied ranks is equal to the average of the ranks.

Test Statistic

This section describes the steps to be followed for calculating the test statistic, H.

The formula for Kruskal-Wallis' H is ...

$$H = \frac{12}{N*(N+1)} * \left(\frac{R_1^2}{n_1} + \frac{R_2^2}{n_2} + ... + \frac{R_k^2}{n_k}\right) - 3*(N+1)$$

... where H is the Kruskal Wallis' test statistic, N is the total number of values, R is the sum of the ranks for each group, n is the number of values for each group, and k is the number of groups.

For the sample data, the test statistic, H, 10.64:

$$((12/(15*(15+1))*((62^2/5)+(16^2/5)+(42^2/5))-(3*(15+1))=10.64$$

Degrees of Freedom

For the sample data, the degrees of freedom (DF) are 2:

$$3-1=2$$

Critical Value and Confidence Level

Recall that if the test shows that the calculated test statistic is lower than the critical value (CV), then one fails to reject —that is, accepts— the null hypothesis. Conversely, if the test shows that the calculated test statistic is higher than the CV, then, one rejects the null hypothesis.

The chi-square table is used for finding the critical values (CVs) of chi-square that correspond to a selected level of confidence, where CL=100%-α, which, for this example, 95% is selected. Table 46.3 shows that the chi-square statistics that correspond to the 95% confidence level, which is CL=100%-5%, is 5.99. The confidence level shows the probability, which is 95%, that the chi-square statistics are less than 5.99.

Table 46.2			
Degrees of Freedom	**Probability (Right-Tail)**		
	0.100	0.050	0.025	0.001
3	4.61	5.99	7.38	9.21

Inference

The calculated test statistic, H, of 10.64 is higher than the critical values (CV) of 5.99, therefore, one rejects H_0.

Lowry

Figure 46.2 shows the output of Lowry's Kruskal-Wallis Test Calculator. [1]

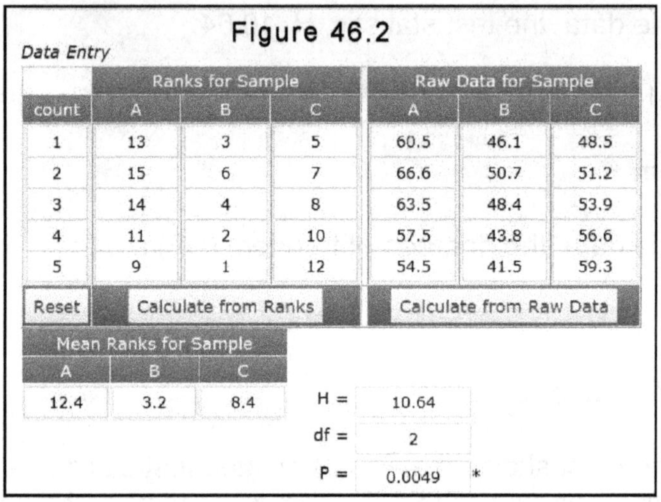

Figure 46.2

Data Entry

count	Ranks for Sample				Raw Data for Sample		
	A	B	C		A	B	C
1	13	3	5		60.5	46.1	48.5
2	15	6	7		66.6	50.7	51.2
3	14	4	8		63.5	48.4	53.9
4	11	2	10		57.5	43.8	56.6
5	9	1	12		54.5	41.5	59.3

Reset	Calculate from Ranks	Calculate from Raw Data

Mean Ranks for Sample		
A	B	C
12.4	3.2	8.4

H = 10.64

df = 2

P = 0.0049 *

Endnote

[1] Lowry, R. (2015). *Kruskal-Wallis test calculator*. URL: http://http://vassarstats.net/

Index